FIXED POINT THEORY AND ITS APPLICATIONS

ACADEMIC PRESS RAPID MANUSCRIPT REPRODUCTION

Seminar on Fixed Point Theory and its Applications, Dalhousie University, 1975.

FIXED POINT THEORY AND ITS APPLICATIONS

Edited by

SRINIVASA SWAMINATHAN

Department of Mathematics
Dalhousie University
Halifax, Nova Scotia, Canada

ACADEMIC PRESS
NEW YORK SAN FRANCISCO LONDON 1976

A Subsidiary of Harcourt Brace Jovanovich, Publishers

ACADEMIC PRESS, INC.
111 Fifth Avenue, New York, New York 10003

United Kingdom Edition published by
ACADEMIC PRESS, INC. (LONDON) LTD.
24/28 Oval Road, London NW1

LIBRARY OF CONGRESS CATALOG CARD NUMBER: 76–53885

ISBN 0–12–678650–X

PRINTED IN THE UNITED STATES OF AMERICA

CONTENTS

LIST OF PARTICIPANTS

FELIX E. BROWDER, University of Chicago, Chicago, Illinois
KUANG-HO CHEN, University of New Orleans, New Orleans, Louisiana
KIM-PEU CHEW, University of Malaya, Kuala Lumpur, Malayasia
M.M. DAY, University of Illinois, Urbana, Illinois
MICHAEL EDELSTEIN, Dalhousie University, Halifax, Nova Scotia, Canada
D. FIELDHOUSE, University of Guelph, Guelph, Ontario, Canada
P.A. FILLMORE, Dalhousie University, Halifax, Nova Scotia, Canada
G. FOX, University of Montreal, Montreal, Quebec, Canada
A. GRANAS, University of Montreal, Montreal, Quebec, Canada
G.B. GUSTAFSON, University of Utah, Salt Lake City, Utah
R.D. HOLMES, Dalhousie University, Halifax, Nova Scotia, Canada
J. INCIURA, University of Waterloo, Waterloo, Ontario, Canada
D.G. KABE, St. Mary's University, Halifax, Nova Scotia, Canada
L.A. KARLOVITZ, University of Maryland, College Park, Maryland
L. KEENER, Dalhousie University, Halifax, Nova Scotia, Canada
MO TAK KIANG, St. Mary's University, Halifax, Nova Scotia, Canada
W.A. KIRK, University of Iowa, Iowa City, Iowa
V. LAKSHMIKANTAM, University of Texas at Arlington, Arlington, Texas
ANTHONY TO-MING LAU, University of Alberta, Edmonton, Alberta, Canada
S. LEELA, State University of New York at Geneseo, Geneseo, New York
TIEN-YIEN LI, University of Utah, Salt Lake City, Utah

W.R. MANN, University of North Carolina, Chapel Hill, North Carolina
P. MORALES, University of Montreal, Montreal, Quebec, Canada
C.W. NORRIS, Memorial University, St. John's, Newfoundland, Canada
R.C. O'BRIEN, Dalhousie University, Halifax, Nova Scotia, Canada
HEINZ-OTTO PEITGEN, Universität Bonn, Wegelerstrasse, Bonn, Germany
W.V. PETRYSHYN, Rutgers University, New Brunswick, New Jersey
M. RADJABALIPOUR, Dalhousie University, Halifax, Nova Scotia, Canada
H. RADJAVI, Dalhousie University, Halifax, Nova Scotia, Canada
J. REINERMANN, University of Aachen, Aachen, Germany
H. SAYEKI, University of Montreal, Montreal, Quebec, Canada
R. SCHÖNEBERG, University of Aachen, Aachen, Germany
S.P. SINGH, Memorial University, St. John's, Newfoundland, Canada
S. SWAMINATHAN, Dalhousie University, Halifax, Nova Scotia, Canada
KOK-KEONG TAN, Dalhousie University, Halifax, Nova Scotia, Canada
A.C. THOMPSON, Dalhousie University, Halifax, Nova Scotia, Canada
CHI SONG WONG, University of Windsor, Windsor, Ontario, Canada
T. J. WOROSZ, Wabash College, Crawfordsville, Indiana

SEMINAR ON FIXED POINT THEORY AND ITS APPLICATIONS

The academic program of the seminar consisted of seven one-hour and fourteen half-hour talks besides a session on problems. The sessions were chaired by: F.E. Browder, M.M. Day, M. Edelstein, A. Granas, Les Karlovitz, W.A. Kirk, W. Petryshyn and A.C. Thompson.

The following talks were given:

Felix E. Browder	(1)	Fixed point theory and mapping theory for multivalued mappings.
	(2)	Maximal monotone operators in non-reflexive Banach spaces.
Kuang-Ho Chen		Semigroups and fixed point theory in biomathematical models.
M.M. Day		Invariant renorming.
Michael Edelstein		Remarks on non-expansive mappings.
A. Granas	(1)	On a method of Poincaré.
	(2)	Some general fixed point and coincidence theorems.
R.D. Holmes	(1)	Linearisation of mappings.
	(2)	A theorem on local radial contractions.

D.G. Kabe		Fixed point theorems in statistical metric spaces.
L.A. Karlovitz		Fixed point theorems of non-expansive mappings and related geometrical considerations.
Mo Tak Kiang		Fixed point theorems of semigroups of self-mappings.
W.A. Kirk		Caristi's fixed point theorem and theory of normal solvability.
V. Lakshmikantam		Fixed points of Lyapounov monotone operators in Banach spaces.
Anthony To-Ming Lau		Some fixed point theorems and their applications to W*-algebras.
C.W. Norris		Some fixed point theorems in locally convex spaces.
W.V. Petryshyn		A-properness and maps of monotone type with applications to elliptic equations.
J. Reinermann	(1)	Fixed point theorems for compact and non-expansive mappings on star-shaped domains.
	(2)	Fixed point theorems of Leray-Schauder type for non-expansive mappings in Hilbert spaces.
S.P. Singh		A fixed point theorem for sum of two mappings.
Chi Song Wong		Some topological and geometrical problems of contractive type.

PREFACE

This volume contains the proceedings of the seminar on Fixed Point Theory and Its Applications held at Dalhousie University, Halifax, Nova Scotia, June 9–12, 1975. The seminar was sponsored by the Canadian Mathematical Congress (as part of the activities of the eastern branch of its summer research institute) and Dalhousie University. The aim of the seminar was to feature recent advances in theoretical and practical aspects of fixed point theory. The geometrical aspects of the theory were emphasized, although the topological and analytical aspects were not entirely ignored.

The seminar was opened with words of welcome by Dr. Henry Hicks, President of Dalhousie University, and introductory remarks by Dr. A.J. Tingley, Professor of Mathematics and Registrar of the University.

The contributions in this volume consist of papers submitted by the participants and by those who intended but were unable to attend the seminar. These papers are valuable not only for their content but also for their representation of current research. They are arranged alphabetically according to author.

I take this opportunity to thank Professors M. Edelstein and A.C. Thompson for their help and advice in the organization of the seminar and in the preparation of this volume; Professor Les Karlovitz for his advice on various matters; Miss Gretchen G. Smith and Miss Paula A. Flemming for secretarial help during the seminar and for typing the manuscript; the Canadian Mathematical Congress and Dalhousie University for financial support for organizing the seminar; and Academic Press for publishing these proceedings.

FIXED POINT THEORY AND ITS APPLICATIONS

FIXED POINT THEOREMS FOR MULTIVALUED MAPPINGS WITH COMPACT ATTRACTORS

Felix E. Browder

In two recent papers ([3],[4]), the writer has established fixed point theorems for locally compact self-mappings f of a large class of smooth spaces X such that f has a compact attractor A, i.e., such that for each point x of X and each neighborhood V of A, there exists an integer $n = n(x,V)$ such that $f^n(x)$ lies in V. It is our purpose in the present paper to extend these results to a broad class of multivalued mappings f (i.e. maps of X into 2^X).

Our principal result for the multivalued case is the following:

<u>Theorem 1</u>. Let X be an open convex subset of a Banach space B, f an upper-semi-continuous mapping of X into $K(X)$, where $K(X)$ denotes the class of non-empty compact convex subsets of X. Suppose that f is locally compact (i.e. for each x in X, there exists a neighborhood U of x in X such that $f(U)$ is relatively compact in X). Suppose that f has a compact attractor in X, i.e., a compact subset A of X such that for each x in X and each neighborhood V of A, there exists an integer $n = n(x,V)$ such that $f^n(x) \subset V$.

Then f has a fixed point x_0 in A, i.e., a point x_0 such that $x_0 \in f(x_0)$.

As a corollary of Theorem 1, we have a theorem of an older type:

<u>Theorem 2</u>. Let X be an open convex subset of a Banach space B, f an upper-semi-continuous, locally compact mapping of X into $K(X)$. Suppose that

$$C_\infty = \bigcap_{j\geq 0} f^j(X)$$

is non-empty and $cl(C_\infty)$ is compact. Suppose further that the orbit of each point of X under f is relatively compact in X.

Then f has a fixed point in C_∞.

The result corresponding to Theorem 2 for the single-valued case was established in [3] and generalized to the case of f a locally condensing mapping in [8]. The proofs given for the single-valued cases of Theorems 1 and 2 rest upon the Lefschetz fixed point theorem for compact self-mappings and yield an extension of that theorem to mappings with compact attractors ([4]). The corresponding proof for Theorem 2 which has been given in [5], for example, depends upon an extension of the Lefschetz fixed point theorem to multivalued maps which is functorial so that the iterates of the mapping f lie in the category of mappings to which the Lefschetz theorem is to apply. Such categories are of a relatively abstract and untransparent character (except to specialists in algebraic topology). We have therefore thought it useful to give a proof of Theorem 1 and thereby of Theorem 2, which does not depend upon a direct generalization of the Lefschetz theorem of this kind. The technique applied in the argument gives a relatively simple extension of the Lefschetz fixed point theorem to a class of multivalued mappings defined by simple geometrical properties. This generalization is only sketched in the presentation below, and will be considered in detail elsewhere.

Section 1. Before we begin the preparation of the proof of Theorem 1, we give the easy proof that Theorem 2 is a consequence of Theorem 1.

Proof of Theorem 2 assuming Theorem 1. Suppose that the hypotheses of Theorem 2 hold, and set $A = cl(C_\infty)$. We shall show that A is a compact attractor for the locally compact mapping f in the sense of Theorem 1.

Let V be an open neighborhood of A in X, and let x be a given point of X. Set

$$Q_n = cl(\bigcup_{j \geq n} f^j(x)) .$$

By hypothesis, each Q_n is a compact subset of X, and the sequence of sets $\{Q_n\}$ is decreasing in n. We assert that for some n, Q_n is contained in V. Indeed, suppose not. Then for each n,

$$Q_n \cap (X \backslash V) \neq \emptyset .$$

Hence, the compact sets $Q_n \cap (X \backslash V)$ have a non-empty intersection. Thus, if we set $Q = \bigcap_{n \geq 0} Q_n$, the compact set Q has a point in common with $X \backslash V$.

On the other hand, we assert that Q is a subset of C_∞ and therefore entirely contained in V. Indeed, let u be a point of Q, and let r be a positive integer. By the definition of Q, we may find a sequence $\{n_j\}$ of positive integers tending to infinity with j and for each j, an element w_j of $f^{n_j}(x)$ such that w_j converges to u as $j \to \infty$. By the definition of $f^{n_j}(x)$, there exists a point y_j in $f^{n_j - r}(x)$ such that $w_j \in f^r(y_j)$. Using the fact that the orbit of x under f is relatively compact in X and passing to an infinite subsequence of the original sequence $\{n_j\}$, we may assume that y_j converges to some y in X as $j \to \infty$. It follows from the upper-semp-continuity of the multi-valued mapping f that $u \in f^r(y)$, i.e., u lies in $f^r(X)$. Since, this is true for each $r > 0$, Q is contained in C.

This contradiction proves that $A = \mathrm{cl}(C_\infty)$ is a compact attractor for the mapping f. Applying Theorem 1, we see therefore that f has a fixed point x_0 in X. It follows immediately from the definition of C_∞ that x_0 must lie in C_∞.

We now begin the preliminary steps to the proof of Theorem 1. These steps of an elementary point-set topological character are essentially analogous to the corresponding steps in the single-valued case with only minor modifications.

Lemma 1. Under the hypotheses of Theorem 1, there exists a compact subset A_1 of X which contains A, is invariant under f, and has the property that there exists a compact convex subset A_0 of A_1 such that for any compact subset K of X, we may find an integer $n(K)$ such that $f^{n(K)}(K) \subset A_0$.

Proof of Lemma 1. By hypothesis, f is locally compact. Hence, we may find a neighborhood U of A in X such that $\mathrm{cl}(f(U))$ is compact (where cl denotes as usual the closure in X). Let $K_0 = A \cup \mathrm{cl}(f(U))$. Since f and each of its iterates is upper semi-continuous and f is locally compact, it follows immediately that for each integer j , $f^j(K_0)$ is a compact subset of X . For each point x of K_0 , there exists a neighborhood U_x of x in X and an integer n_x such that $f^{n_x}(U_x) \subset U$. The compact set K_0 can be covered by a finite subfamily of such neighborhoods $\{U_{x_1},\ldots,U_{x_r}\}$. Let $m = \max\{n_{x_j} : 1 \le j \le r\}$. Then we set

$$A_1 = \bigcup_{j=0}^{m} f^j(K_0),$$

The set A_1 is compact since it is the union of a finite number of compact sets. It contains A since K_0 contains A . We assert that A_1 is invariant under f . We note that since $f(f^j(K_0)) = f^{j+1}(K_0)$, it suffices to prove that $f^{m+1}(K_0) \subset A_1$. Let x be any point of K_0 . Then x lies in some U_{x_j} and $f^{n_j}(x) \subset U$. In particular,

$$f^{n_j+1}(x) \subset f(U) \subset K_0 .$$

Thus,

$$f^{m+1}(x) = f^{m-n_j}(f^{n_j+1}(x)) \subset f^{m-n_j}(K_0) \subset A_1 .$$

We now assert that A_1 has the property that for each compact set K , there exists an integer $n(K)$ such that $f^n(K) \subset A_1$ for $n \ge n(K)$. For each point u of K , there exists an integer n_u and a neighborhood U_u in X such that $f^{n_u}(U_u) \subset U$, and hence $f^{n_u+1}(U_u) \subset A_1$. Taking a finite covering of K by such neighborhoods U_{u_k} and letting $n(K)$ be the maximum of $\{n_{u_k}+1\}$, it follows immediately that for $n \ge n(K)$, $f^n(K) \subset A_1$.

Finally, such a compact invariant subset A_1 may be obtained for each attractor A . We start with the original attractor A , construct the corresponding A_1 as above, and form a new attractor A' as the convex closure in X of the compact set A_1 . A' is compact in B and contained in X , and hence is compact in X . We now form the corresponding invariant set $(A')_1$ by the above process. The resulting set (which we can designate as the new A_1) will then have the required properties of the conclusion of the lemma.

q.e.d.

Lemma 2. Under the hypotheses of Theorem 1, let A_1 be a compact subset of X which is invariant under f and satisfying the conclusions of Lemma 1. Then there exists an open neighborhood U_1 of A_1 such that $f(U_1) \subset U_1$ and $f(U_1)$ has compact closure in X.

<u>Proof of Lemma 2</u>. Since f is locally compact, there exists an open neighborhood U_0 of A_1 such that $cl(f(U_0)) = K_0$ is a compact subset of X. By Lemma 1, there exists an integer n such that $f^n(K_0) \subset A_1$.

Let

$$U_1 = \{x | x \in U_0 , f^j(x) \subset U_0 \text{ for } 1 \leq j \leq n\} .$$

A_1 is obviously contained in U_1 and by the upper-semi-continuity of f, U_1 is an open neighborhood of A_1 in X. Moreover, $f(U_1) \subset f(U_0)$ so that $cl(f(U_1))$ is compact in X. It therefore suffices to show that U_1 is invariant under f.

Suppose that x lies in U_1. Then for any u in $f(x)$, $f^k(u) \subset U_0$ for $0 \leq k \leq n-1$. On the other hand,

$$f^n(u) \subset f^{n+1}(x) \subset f^{n+1}(U_0) \subset f^n(K_0) \subset A_1 \subset U_0 .$$

Hence, u lies in U_1, and U_1 is invariant under f. q.e.d.

<u>Lemma 3</u>. Suppose under the hypotheses of Theorem 1 that A_1 is a compact subset of X with $f(A_1) \subset A_1$ which satisfies the conclusions of Lemma 1 and that U_1 is a neighborhood of A_1 in X which satisfies the conclusions of Lemma 2. Then there exists an open neighborhood U_2 of A_1 in U_1 which is mapped into itself by f such that $cl(f(U_2))$ is a compact subset of U_2 while for some integer n, $f^n(U_2)$ is contained in the compact convex subset A_0 of A_1.

<u>Proof of Lemma 3</u>. Since $K_0 = cl(f(U_1))$ is a compact subset of X, it follows from Lemma 1 that there exists an integer n such that $f^{n-1}(K_0) \subset A_0$. For this particular integer n, $f^n(U_1) \subset A_0$, so that for any subset U_2 of U_1, we have $f^n(U_2) \subset A_0$.

Having obtained n as above, we begin the construction of the neighborhood U_2 of A_1 by choosing a small neighborhood V of A_1 in U_1 such that $f^{j-1}(cl\ f(V)) \subset U_1$ for $1 \leq j \leq n-1$. Starting with V, we now construct a sequence

$$\{V_1, \ldots, V_n\}$$

of open neighborhoods of A_1 in U_1 which has the following properties:

(a) $V_n = V$.

(b) $cl(f(V_j)) \subset V_{j+1}$, $(1 \le j \le n-1)$.

(c) $f^{j-1}(cl(f(V))) \subset V_j$, $(1 \le j \le n-1)$.

Note that if such a construction is possible, then

$$U_2 = \bigcup_{j=1}^{n} V_j$$

will satisfy the conclusions of Lemma 3.

We begin with the construction of V_{n-1} . By the regularity of the topological space X , we can find an open neighborhood W_{n-1} of A_1 such that $cl(W_{n-1}) \subset V_n = V$. For each x in $f^{n-2}(cl(f(V)))$, we note that x lies in $f^{n-2}(K_0)$ with $K_0 = cl(f(U_1))$ so that $f(x) \subset A_1$. Hence, there exists a neighborhood U_x of x such that $f(U_x) \subset W_{n-1}$. We set

$$V_{n-1} = \bigcup_{x \in f^{n-2}(cl(f(V)))} U_x .$$

Then $f(V_{n-1}) \subset W_{n-1}$, so that $cl(f(V_{n-1})) \subset V_n$. On the other hand, by its construction, $f^{n-2}(cl(f(V))) \subset V_{n-1}$.

Suppose now that we have constructed a sequence $\{V_r, V_{r+1}, \ldots, V_n\}$ satisfying conditions (a), (b) and (c) for $j \ge r$. If $r > 1$, we shall now construct V_{r-1} so that these properties hold for $j \ge r-1$. By the inductive assumption

$$f^{r-1}(cl(f(V))) \subset V_r .$$

Since $cl(f(V))$ is compact, $f^{r-2}(cl(f(V))$ is compact as well as $f^{r-1}(cl(f(V)))$. We choose an open neighborhood W_r of $f^{r-1}(cl(f(V)))$ in V_r so that $cl(W_r) \subset V_r$. For each x in $f^{r-2}(cl(f(V)))$, $f(x) \subset W_r$. Hence, each such x has an open neighborhood U_x such that $f(U_x) \subset W_r$. We set

$$V_{r-1} = \bigcup_{x \in f^{r-2}(cl(f(V)))} U_x .$$

Then V_{r-1} contains $f^{r-2}(cl(f(V)))$, while $f(V_{r-1}) \subset W_r$. Hence

$$cl(f(V_{r-1})) \subset cl(W_r) \subset V_r .$$

Thus, the sequence $\{V_r, \ldots, V_n\}$ has been extended by one step, and the proof of the lemma is complete. q.e.d.

Section 2. To complete the proof of Theorem 1, it suffices to apply the Lefschetz fixed point theorem to the compact mapping $g = f|_{U_2}$ of the open subset U_2 of X into $K(U_2)$. To avoid introducing the relatively sophisticated form the Lefschetz fixed point theorem must take in order to be able to be applied to iterates of the mapping g, we shall take another, more elementary path to the same conclusion which uses the much simpler result in the single-valued case.

We employ a simple method of approximating multivalued maps by single-valued mappings along lines introduced by Cellina and Lasota in [6].

Definition. Let X and Y be topological spaces, f a mapping of X into 2^Y, β an open covering of X, ξ an open covering of Y. Then a continuous mapping h of X into Y is said to be a (β,ξ) approximation for f if for each x in X, there exists u in X, w in $f(u)$ such that the pair $\{x,u\}$ lies in a common element of the covering β and the pair $\{h(x),w\}$ lies in a common element of the covering ξ.

Lemma 4. Let X be an open subset of a Banach space B, f an upper-semicontinuous mapping of X into $K(X)$. Then if β is a covering of X, ξ a covering of X, there exists a continuous mapping h of X into B which is a (β,ξ) approximation for f.

Proof of Lemma 4. Since X is paracompact and since the conditions for an approximation become stronger if the pair of coverings $\{\beta,\xi\}$ is replaced by a pair of finer coverings, we may assume without loss of generality that $\beta=\xi$ and that this covering β is locally finite.

We assert that there exists a locally finite covering β_0 of X such that β_0 is a star-refinement of β, i.e., if $\{V_\gamma\}$ is a finite collection of elements of the covering β_0 whose intersection is non-empty, then $\cup_\gamma V_\gamma$ is contained in some element of β. Since every open covering of X has a locally finite refinement, it suffices to prove that some covering β_0 exists which is a star-refinement of β without specification of the local finiteness of β_0. Let N_β be the nerve of the covering , i.e., N_β is a simplicial complex whose vertices correspond one-to-one to the elements of the covering β with a finite collection of vertices spanning a simplex if and only if the corresponding elements of the covering β have a non-empty intersection. Let $\{q_\delta\}$ be a partition of unity subordinated to the covering β. Then there exists a continuous mapping ζ of X into N_β such that $\zeta(x)$ is the point of the simplex spanned by the elements $\{V_1,\ldots,V_r\}$ of which contain x with the barycentric coordinates $\{q_{V_1}(x),\ldots,q_{V_r}(x))\}$. N_β is given the topology such that a subset is open if and only if its intersection with every finite simplicial subcomplex of N is open. In this topology, the star of each vertex $\{p_V\}$ is open, and its inverse image under ζ is the open set $\{x | q_V(x) > 0\}$. If we take the first barycentric subdivision of N_β, the covering by stars of the barycentric subdivision is a star-refinement of the covering by stars of vertices in N_β. The inverse image of this refinement under ζ is the desired star-refinement β_0 of β.

For each u in X, f(u) is assumed to be compact, non-empty, and convex. It follows from its compactness that there exists a constant d(u) such that any point w in the d(u)-neighborhood of f(u) must lie in the same element of the β-covering as a point w_0 of f(u). By the upper-semi-continuity of f, for each u in X, there exists an open neighborhood U_u of u in X such that for x in U_u, f(x) is contained in the d(u)-neighborhood of f(u). We may assume that each U_u is contained in some element of the covering β. Let β_1 be a locally finite covering of X which is a star-refinement of the covering by the sets U_u. For each V_δ in β_1, let u_δ be a point in V_δ and let w_δ be a point in $f(u_\delta)$. Form a partition of unity $\{q_\delta\}$ subordinated to the covering β_1. If we set

$$h(x) = \Sigma_\delta\, q_\delta(x) w_\delta\,,$$

the sum makes sense since only a finite number of terms differ from zero in the neighborhood of a given point. The mapping h is therefore a continuous mapping h of X into B. We shall show that it is a (β,β)-approximation to our original multivalued mapping f.

For a given x in X, let $\{V_{\delta_1},\ldots,V_{\delta_r}\}$ be the finite family of elements of β_1 which contain the point x. Since β_1 is a star-refinement of the covering $\{U_u : u \in X\}$ and since the intersection of the sets V_{δ_j} contains the point x, there exists u in X such that all the sets V_{δ_j} are contained in U_u. In particular, u_{δ_j} lies in U_u for each j, and hence w_{δ_j} lies in the d(u)-neighborhood of f(u). Since h(x) is a convex linear combination of the points w_{δ_j} and since f(u) is convex, it follows that h(x) lies in the d(u)-neighborhood of $f(u)$, i.e. there exists an element w of $f(u)$ such that $h(x)$ and w lie in the same element of β. On the other hand, x lies in U_u which is contained in some element of β. Hence, x and u lie in a common element of β. q.e.d.

Lemma 5. Let X be an open subset of the Banach space B, f an upper-semi-continuous mapping of X into $K(X)$ such that $cl(f(X))$ is a compact subset of X. Then for each open covering β of X, there exists a continuous mapping h of X into X with $cl(h(X))$ compact in X such that h is a (β,β) approximation to f.

Proof of Lemma 5. If we consider the approximation h which was constructed in the proof of Lemma 4, we note that the values taken on by h lie in the convex closure of the set $f(X)$. If the latter set is relatively compact in B, h will be a compact mapping of X into B. On the other hand, if $cl(f(X))$ is a compact subset of X, its distance from the complement of X in B is a positive number d. Since $h(x)$ for every x will lie in the d_1-neighborhood of $f(X)$ with $d_1 < d/2$, if β is taken sufficiently fine, $h(X)$ will have distance at least $d/2$ from $B\backslash X$ for such approximations. Hence $cl(h(X))$ will be a subset of X. q.e.d.

Lemma 6. Let X be an open subset of the Banach space B, f an upper-semi-continuous mapping of X into $K(X)$ such that $cl(f(X))$ is a compact subset of X. Let β be a given covering of X, n a positive integer. Then there exists β_2 such that if h is a (β_2,β_2)-approximation to f, then h is a compact mapping of X into X and h^n is a (β,β)-approximation to f^n.

Proof of Lemma 6. It suffices to prove that if f and g are mappings of X into 2^X and if the covering β is given, then there exists an open covering β_2 of X such that if h is a (β_2,β_2)-approximation of f and if k is a (β_2,β_2)-approximation to g, then hk is a (β,β)-approximation to fg, (β_2 of course depending upon f and g).

By the concept of approximant in the sense of our definition, each x in X, there exists u in X, w in $g(u)$ such that

$\{u,x\}$ lies in an element of β_2,

$\{w,k(x)\}$ lies in an element of β_2.

Similarly, if we set $y = k(x)$, there exists v in X , z in $f(v)$ such that

$\{v,y\}$ lies in an element of β_2 ,

$\{z,h(y)\}$ lies in an element of β_2 .

Hence,

$\{v,w\}$ lies in a common element of $St(\beta_2)$.

We let β_1 be a star-refinement of β , and assume that β_2 is finer than β_1 and β_2 is sufficiently fine. We may ensure in particular that if $\{v,w\}$ lies in a common element of $St(\beta_2)$, then

$f(v)$ lies in the β_1-neighborhood of $f(w)$.

Then $h(y) = h(k(x))$, which lies in a common element of β_2 with an element of $f(v)$, will lie in a common element of $St(\beta_1)$ with some element of $f(w) = f(g(u))$. Hence hk is a (β,β)-approximation to fg for such a choice of β_2 . q.e.d.

Proof of Theorem 1 completed. Let X be replaced by the open set U_2 , so that f is a compact mapping of X into $K(X)$ such that for a given integer n , $f^n(X)$ is contained in a compact convex subset A_0 of X . Let h be a (β,β)-approximation to f for β sufficiently fine. Then by Lemma 5, h is a compact mapping of X into X . By Lemma 6, if $d > 0$ is given and if β is sufficiently fine, $h^n(X)$ will be contained in the d-neighborhood of A_0 . For $d > 0$ and sufficiently small, $N_d(A_0)$ is a convex subset of X and hence contractible to a point in X . Applying the Lefschetz fixed point theorem to h , it follows that the Lefschetz number of h must equal $+1$, and hence h has a fixed point x_β in X . Since x_β lies in the $St(\beta)$-neighborhood of $f(x_\beta)$, it follows by taking a sequence with $mesh(\beta_j) \to 0$, that we can choose a sequence x_{β_j} converging to x and $x \in f(x)$. q.e.d.

References

[1] Browder, F.E., On a generalization of the Schauder fixed point theorem, Duke Math. Jour., 26 (1959), 291-303.

[2] __________, Asymptotic fixed point theorems, Math. Annalen, 185 (1970), 38-60.

[3] __________, Some new asymptotic fixed point theorems, Proc. Nat. Acad. Sci., 71 (1974), 2734-2735.

[4] __________, The Lefschetz fixed point theorem and asymptotic fixed point theorems, Partial Differential Equations and Related Topics, Springer Lecture Notes in Math., Vol. 446 (1975), 96-122.

[5] Calvert, B., The local fixed point index for multivalued transformations in a Banach space, Math. Annalen, 190 (1970), 119-128.

[6] Cellina, A. and A. Lasota, A new approach to the theory of the topological degree for multivalued mappings, Accad. Naz. Lincei rend. Cl. Sci. Fiz. Mat. Natur., 47 (1969), 434-440.

[7] Granas, A., The Leray-Schauder degree and the fixed point theory for arbitrary ANR's, Bull. Soc. Math. France, 100 (1972), 209-228.

[8] Nussbaum, R., Asymptotic fixed point theorems for local condensing mappings, Math. Annalen, 191 (1971), 181-195.

CONTINUATION OF SOLUTIONS OF EQUATIONS UNDER HOMOTOPIES OF SINGLE-VALUED AND MULTIVALUED MAPPINGS

Felix E. Browder

In a recent paper ([6]), the writer established results on continuation of solutions of equations involving mappings which satisfy the conditions of the theory of normal solvability. The argument was based upon a simple general principle of a fundamentally topological nature for continuation of solutions of equations under homotopy. In the present paper, we shall sharpen and extend these results for single-valued mappings as well as give a generalization to the multivalued case.

Let X be a set, Y a metric space, f a mapping of X into Y. We distinguish a subset X_0 of X as a sort of abstract "boundary" and impose hypotheses separately upon the behavior of f on the points of X_0 and of $X \setminus X_0$.

Definition 1. The mapping f is said to be full on $\{X, X_0\}$ if each connected set C of $Y \setminus \mathrm{cl}(f(X_0))$ which meets $f(X)$ must lie entirely in $f(X)$.

Definition 2. Let $\{f_t: 0 \le t \le 1\}$ be a one parameter family of mappings of X into Y. Then the family is said to be a uniform homotopy if for each t_0 in $[0,1]$ and each $\beta > 0$, there exists $\delta > 0$ such that if $|t-t_0| < \delta$, then

$$d(f_t(x), f_{t_0}(x)) < \beta$$

for all x in X.

Note that on principle, the set X has no topological structure, and no structural hypotheses except fullness are imposed upon mappings.

Theorem 1. Let X be a set, Y a metric space which is locally connected, $\{f_t: 0 \le t \le 1\}$ a uniform homotopy of mappings of X into

Y. Let X_0 be a subset of X, and suppose that each f_t is full with respect to the pair (X,X_0). Suppose further that for each t in $[0,1]$, $f_t(X)$ is closed in Y.

Let $\{y(t): 0 \leq t \leq 1\}$ be a continuous curve in Y such that $y(0)$ lies in $f(X)\backslash cl(f_0(X_0))$. Suppose that for each t in $[0,1]$, $y(t)$ does not lie in $cl(f_t(X_0))$. Then for each t in $[0,1]$, $y(t)$ lies in $f_t(X)$.

Proof of Theorem 1. Let $Q = \{t | t \in [0,1], y(t) \in f_t(X)\}$. To prove the theorem means to prove that $Q = [0,1]$. By the assumption on $y(0)$, 0 lies in Q, and Q is therefore non-empty. Since $[0,1]$ is connected, it suffices to prove that Q is both open and closed in $[0,1]$.

Proof that Q is closed in $[0,1]$: Let $\{t_j\}$ be a sequence in Q converging to t in $[0,1]$. For each j, there exists a point x_j in X such that $f_{t_j}(x_j) = y(t_j)$. Since $\{y(t): t \in [0,1]\}$ is a continuous curve, $d(y(t_j),y(t)) \to 0$ as $j \to \infty$. By the condition that $\{f_t\}$ is a uniform homotopy,

$$d(f_{t_j}(x_j),f_t(x_j)) \to 0 , \qquad (j \to +\infty) .$$

Hence

$$d(f_t(x_j),y(t)) \to 0 , \qquad (j \to +\infty) ,$$

i.e., $y(t)$ lies in the closure in Y of $f_t(X)$. Since $f_t(X)$ is assumed to be closed in Y, it follows that $y(t)$ lies in $f_t(X)$. Hence t lies in Q, and Q has thereby been shown to be closed.

Proof that Q is open in $[0,1]$: Suppose that t_0 lies in Q. We seek to show that there exists $\delta > 0$ such that for $|t-t_0| < \delta$, t lies in Q, i.e., $y(t)$ lies in $f_t(X)$.

By assumption, there exists u in X such that $f_{t_0}(u) = y(t_0)$. Since $y(t_0)$ lies in $Y\backslash cl(f(X_0))$, there exists a ball $B_{3r_0}(y(t_0))$ around $y(t_0)$ in Y with $r_0 > 0$ which is disjoint from $f_{t_0}(X_0)$. Since $\{f_t\}$ is a uniform homotopy, there exists $\delta > 0$ such that for all x in X, $d(f_t(x),f_{t_0}(x)) < r_0$ if $|t-t_0| < \delta$. In particular, each point of $f_t(X_0)$ for such t lies at distance at most r_0 from $f_{t_0}(X_0)$ and hence outside the ball $B_{2r_0}(y(t_0))$. Hence $cl(f_t(X_0))$

for each such t lies in the complement of $B_{2r_0}(y(t_0))$. On the other hand, since Y is locally connected, there exists a connected neighborhood U_0 of $y(t_0)$ in Y which is contained in $B_{r_0}(y(t_0))$. By making $\delta > 0$ still smaller, we may ensure that for $|t-t_0| < \delta$, $f_t(u)$ lies in U_0 and $y(t) \in U_0$.

By hypothesis, each f_t is full with respect to the pair (X,X_0). The connected set U_0 does not meet $cl(f_t(X_0))$ for $|t-t_0| < \delta$ while it contains the point $f_t(u)$. Hence by the property of fullness, U_0 is contained in $f_t(X)$ for such t . On the other hand, $y(t)$ lies in U_0 for $|t-t_0| < \delta$. Hence, $y(t)$ lies in $f_t(X)$ for $|t-t_0| < \delta$, i.e., t lies in Q . Thus Q is open in $[0,1]$.

Proof of Theorem 1 concluded: Since $[0,1]$ is connected and Q is nonempty while being both open and closed in $[0,1]$, Q must be the whole of $[0,1]$. q.e.d.

The property of fullness is satisfied as a consequence of the following hypothesis upon the range of f (together with the assumption that $f(X)$ is closed in Y), as we show in detail in Theorem 2 below:

Definition 3. Let Y be a Banach space, S a subset of Y . Then S is said to be infinitesimally full in Y if the following condition is satisfied:

(s): For each y in $Y \cap S$, there exists $r > 0$, $p < 1$ such that $B_r(y)$ intersects S and for each v in $S \cap B_r(y)$, there exists a sequence $\{v_j\}$ in $S\backslash\{v\}$ converging to v and a sequence of positive real numbers $\{\xi_j\}$ such that

$$||\xi_j(v_j-v)-y|| \leq p||v-y|| .$$

Theorem 2. Let X be a set, Y a Banach space, f a mapping of X into Y such that $f(X)$ is closed in Y . Let X_0 be a subset of X , and suppose that $f(X\backslash X_0)$ is infinitesimally full in Y .

Then f is full with respect to the pair (X,X_0) .

Proof of Theorem 2. We use the following basic result from the theory of normal solvability ([1],[4]):

Proposition. Let S be a closed subset of the Banach space Y , y a point of $Y \cap S$, $r > 0$, $p < 1$ such that $B_r(y) \cap S \neq \emptyset$. Then there exists a point u of $S \cap B_r(y)$ and $\delta > 0$ such that the conical segment

$$V = \{v: v = u + \xi z \ , \ 0 < \xi < \delta \ , \ ||z-y|| < p||u-y||\}$$

has a void intersection with S .

<u>Proof of Theorem 2 continued</u>: Let C be a connected component of $Y \backslash cl(f(X_0))$ such that C contains a point of $f(X)$. We wish to show that C is contained in $f(X)$. Since C is connected, it suffices to show that $f(X) \cap C$ is open and closed in C . Since $f(X)$ is closed in Y , it is obvious that $f(X) \cap C$ is closed in C . It therefore suffices to show that $f(X) \cap C$ is open in C .

Let y_0 be a point of $f(X) \cap C$. Since C is disjoint from $cl(f(X_0))$, there exists $r > 0$ such that $B_{3r}(y_0)$ is disjoint from $f(X_0)$. Let y be any interior point of $B_r(y_0)$. Then $B_r(y) \cap S$ is non-empty where $S = f(X)$, and we may apply the Proposition stated above to the closed set S in Y . If y does not lie in $f(X)$, we may find a conical segment as above which is disjoint from $f(X) = S$. By the fact that $B_r(y) \cap f(X) = B_r(y) \cap f(X_0)$, it follows from the property (s) of Definition 3 that if $f(X_0)$ is infinitesimally full, each such conical segment V at any point u of $f(X) \cap B_r(y)$ must contain points of $f(X)$. Hence y_0 lies in the interior of $f(X)$, i.e., $f(X) \cap C$ is open in C . q.e.d.

Obviously, the condition of infinitesimal fullness is a drastic weakening of more classical conditions like the condition that f is open at each point of $X \backslash X_0$ or that f is differentiable at each point of $X \backslash X_0$ and that its differential df_u at each point u of $X \backslash X_0$ has dense range.

We now turn to modified forms of Theorem 1 which parallel corresponding forms of the continuation property for the case of mappings of the form (I-C) with non-zero degree for C compact in Banach spaces.

<u>Theorem 3</u>. Let X be a set, Y a locally connected metric space, W a subset of $X \times [0,1]$, W_0 a subset of W , F a mapping of W into Y . For each t in $[0,1]$, let $X_t = \{x | x \in X, [x,t] \in W\}$, $X_{0,t} = \{x | x \in X, [x,t] \in W_0\}$. Let $y(t): 0 \le t \le 1$ be a continuous curve in Y . Suppose that the following conditions all hold:

(a) If f_t is the map of X_t into Y given by $f_t(x) = F(x,t)$, then f_t is full with respect to the pair $(X_t, X_{0,t})$.

(b) For each t in $[0,1]$, $y(t)$ lies in $Y \backslash cl(f_t(X_{0,t}))$, while for some t_0, $y(t_0)$ lies in $f_{t_0}(X_{t_0} \backslash X_{t_0}, 0)$.

(c) If for a sequence $\{t_j\}$ converging to t, there exists a sequence $\{x_j\}$ with x_j in X_{t_j} such that $f_{t_j}(x_j)$ converges to y in Y, then y lies in $f_t(X)$. If for a sequence $\{t_j\}$ converging to t in $[0,1]$, we have a sequence $\{x_j\}$ with each x_j in X_{0,t_j} such that $f_{t_j}(x_j) \to y$, then y lies in $cl(f_t(X_{0,t}))$.

(d) For each $[x,t]$ in $W \backslash W_0$, there exists $\gamma > 0$ such that the interval $\{x\} \times [t-\gamma, t+\gamma]$ lies in W, and the mapping F of this interval into Y is continuous.

Then: $y(t)$ lies in $f_t(X_t)$ for each t in $[0,1]$.

Proof of Theorem 3. Let $Q = \{t | t \in [0,1], y(t) \in f_t(X)\}$. Since t_0 lies in Q by condition (b), in order to prove that $Q = [0,1]$ it suffices to prove that Q is both open and closed in $[0,1]$.

Suppose that $\{t_j\}$ is a sequence in Q which converges to t in $[0,1]$. For each j, there exists a point x_j in X_{t_j} such that $f_{t_j}(x_j) = y(t_j)$. Since $y(t)$ is continuous in t, $y(t_j) \to y(t)$ as j goes to infinity. Applying condition (c) of the hypothesis, we see that $y(t)$ lies in $f_t(X)$. i.e., t lies in Q. Hence Q is closed in $[0,1]$.

Suppose on the other hand, that t lies in Q. Then there exists $[x,t]$ in $W \backslash W_0$ such that $f_t(x) = y(t)$. We assert that there exists $r > 0$ and $\delta > 0$ such that for $|t_1 - t| < \delta$, $B_r(y(t)) \cap f_{t_1}(X_{0,t_1}) = \emptyset$. Indeed, suppose not. Then there exists a sequence $\{t_j\}$ converging to t and a sequence $\{x_j\}$ with each x_j in X_{0,t_j} such that $f_{t_j}(x_j)$ has distance less than 2^{-j} from $y(t)$. Applying the second part of condition (c), it follows that $y(t)$ lies in $cl(f_t(X_{0,t}))$ which contradicts condition (a) of the hypothesis. Hence such r and δ exist.

Consider s with $|s-t| < \delta$. Then $cl(f_x(X_{0,s}))$ is disjoint from $B_{r_1}(y(t))$ for $r_1 < r$. Since Y is locally connected, there exists a connected neighborhood U of $y(t)$ contained in $B_{r_1}(y(t))$.

Since $\{y(t): 0 \le t \le 1\}$ is assumed to be a continuous curve, we may choose $\delta > 0$ still smaller so that $y(s)$ lies in U for $|s-t| < \delta$. By condition (d), finally, we may assume $\delta > 0$ still smaller and ensure that the set $\{x\} \times (t-\delta, t+\delta)$ lies in W and that for $|s-t| < \delta$, $f_s(x)$ lies in U_0.

If we collect these facts, we see that for s in the interval $|s-t| < \delta$, the connected set U lies in $Y \backslash cl(f_s(X_{0,s}))$ and contains the point $f_s(x)$ of $f_s(X)$. By condition (a), f_s is full with respect to the pair $(X_s, X_{0,s})$. Hence U is contained in $f_s(X_s)$. Since $y(s)$ lies in U for such s, it follows that $y(s)$ lies in $f_s(X_s)$ and the set Q is open in $[0,1]$.

Thus Q is non-empty, open and closed in $[0,1]$. Thus $Q = [0,1]$. q.e.d.

We may apply Theorem 3 under more precise hypotheses to obtain results on continuity of solutions under the continuation of mappings.

Theorem 4. Let X be a metric space, Y a locally connected metric space, G an open subset of $X \times [0,1]$, W the closure of G in $X \times [0,1]$ and W_0 the boundary of W. Let F be a continuous mapping of W into Y with F proper, $C = \{y(t): 0 \le t \le 1\}$ a continuous curve in Y. Let $X_t = \{x | [x,t] \in W\}$, $X_{0,t} = \{x | [x,t] \in W_0\}$, and let f_t be the mapping of X_t into Y given by $f_t(x) = F(x,t)$. Suppose that:

(a) For each t in $[0,1]$, f_t is full with respect to any pair $(V_t, \partial V_t)$ with $V_t \subset X_t \backslash X_{0,t}$.

(b) The point $y(t_0)$ lies in $f_{t_0}(X_{t_0})$ while for all t in $[0,1]$, $y(t)$ lies in $Y \backslash f_t(X_{0,t})$.

Let $D = \{[x,t] | [x,t] \in W, f_t(x) = y(t)\}$.

Then: Each connected component of D intersects all X_t and therefore passes from the base to the top of the cylindrical set W.

Proof of Theorem 4. Since C is a continuous curve, the corresponding point set which we continue to denote by C is a compact subset of Y. Since F is proper and $D = F^{-1}(C)$, D is a compact subset of W which is disjoint from W_0. Hence D is a compact subset of G.

To establish the assertion of the Theorem, it suffices to show that each connected component D_0 of D intersects both X_0 and X_1. Suppose a given component D_0 does not. Since D is a compact metric space in its induced topology, D_0 is the intersection of a decreasing sequence $\{V_j\}$ of subsets of D which are both open and closed in D. Since each V_t is compact, it follows that if D_0 does not intersect $X \times \{1\}$, all the V_j past some point in the sequence will not intersect $X \times \{1\}$. Choose such an index j, and let

$$D_1 = V_j \text{ , } D_2 = D - V_j \text{ .}$$

D_1 and D_2 are mutually disjoint compact sets and therefore have a positive distance 3δ. We may assume $\delta > 0$ still smaller so that $N_\delta(D_1)$, the δ-neighborhood of D_1 in W, has no points of $X \times \{1\}$ in its closure, and is contained in $W \backslash W_0$.

Let V be the closure of $N_\delta(D_1)$ in W, V_0 its boundary, and consider the mapping F restricted to V. Let $V_t = \{x \mid [x,t] \in V\}$ and $V_{0,t} = \{x \mid [x,t] \in V_0\}$. We assert that the mapping F defined with respect to the pair (V,V_0) satisfies the hypotheses of Theorem 3. Conditions (a) and (b) follow from the hypothesis (a) of Theorem 4, together with the fact that V_0 obviously has no points in common with D. Moreover, F, being proper, is a closed mapping of V into Y, and each mapping f_t is a proper and hence closed mapping of V_t into Y. In particular, $f_t(V_{0,t})$ is closed, so that $y(t)$ does not lie in $cl(f_t(V_{0,t}))$. Condition (d) follows from the continuity of the map F and the fact that V is the closure of an open set. Finally, condition (c) of the hypothesis of Theorem 3 follows from the fact that F is proper. Indeed, suppose $F(x_j,t_j) \to y$. Since a convergent sequence together with its limit forms a compact set in a metric space, the sequence $\{x_j\}$ is relatively compact in X. We may choose a convergent subsequence with limit x. Then $F(x,t) = \lim F(x_j,t_j) = y$, and the appropriate conclusions follow.

If we apply Theorem 3, we obtain the conclusion that V_1 is non-empty, contrary to its construction. This contradiction shows that D_0 must intersect $X \times \{1\}$, and similarly $X \times \{0\}$. Hence, D_0 intersects each X_t. q.e.d.

(II) Multivalued mappings

We now turn to the extension of the preceding results to multi-valued mappings, i.e. mappings from X into 2^Y. This extension demands a concept not present in the single-valued case, that of the regular points of the multivalued mapping T as described in the following definition.

Definition 3. Let X be a set with subset X_0, Y a metric space, T a mapping of X into 2^Y. A point u of X is said to be a regular point of the mapping T with respect to the pair (X,X_0) if there exists $\delta > 0$ such that $N_\delta(T(u)) \backslash N_\delta(T(X_0))$ is non-empty and connected.

We also have the adaptations of the definitions of fullness and of uniform homotopy to the case of multivalued mappings.

Definition 4. Let X be a set, X_0 a subset of X, Y a metric space, T a mapping of X into 2^Y. Then T is said to be full with respect to the pair (X,X_0) if any connected set C in Y which intersects $T(X)$ and does not meet $cl(T(X_0))$ must be entirely contained in $T(X)$.

Definition 5. Let $\{T_t: 0 \le t \le 1\}$ be a family of mappings from X into 2^Y indexed on $[0,1]$. Then $\{T_t\}$ is said to be a uniform homotopy from X into 2^Y if for each t_0 in $[0,1]$ and each $\beta > 0$, there exists $\delta > 0$ such that if $|t-t_0| < \delta$, then $T_t(x) \subset N_\beta(T_{t_0}(x))$ for any x in X.

Theorem 5. Let X be a set, X_0 a subset of X, Y a metric space, $\{T_t: 0 \le t \le 1\}$ a uniform homotopy of mappings of X into 2^Y. Suppose that each T_t is full with respect to the pair (X,X_0) and that for each t in $[0,1]$, $T_t(X)$ is closed in Y. Let $C = \{y(t): 0 \le t \le 1\}$ be a continuous curve in Y such that $y(t)$ does not lie in $cl(T_t(X_0))$ for any t in $[0,1]$, while for each such t, $T_t^{-1}(y(t))$ consists entirely of regular points of T_s with respect to the pair (X,X_0) for s near t.

Suppose finally that $y(0)$ lies in $T_0(X)$.

Then $y(t)$ lies in $T_t(x)$ for all t in $[0,1]$.

Proof of Theorem 5. Let $Q = \{t | t \in [0,1], y(t) \in T_t(X)\}$. Since 0 lies in Q, in order to prove that $Q = [0,1]$ it suffices to show that Q is both open and closed in $[0,1]$.

To show that Q is closed in $[0,1]$, let $\{t_j\}$ be a sequence in Q converging to t in $[0,1]$. For each j , there exists a point x_j in X such that $y(t_j) \in T_{t_j}(x_j)$. Since $\{T_t\}$ is a uniform homotopy, for each $\beta > 0$, there exists an index $j(\beta)$ such that for $j \geq j(\beta)$,

$$y(t_j) \in N_\beta(T_t(x_j)) .$$

Since $y(t_j)$ converges to $y(t)$ as $j \to \infty$, it follows that $y(t)$ lies in the closure of $T_t(X)$. Since $T_t(X)$ is assumed to be closed in Y , $y(t)$ therefore lies in $T_t(X)$ and t therefore lies in Q . Hence, Q is closed in $[0,1]$.

To show that Q is open in $[0,1]$, suppose that t lies in Q . Then $y(t) \in T_t(u)$ and by hypothesis, u is a regular point of T_t with respect to the pair (X,X_0) . Hence, there exists $\delta > 0$ such that

$$C = N_\delta(T_t(u))\backslash N_\delta(T_t(X_0))$$

is a non-empty connected set. If $|s-t|$ is sufficiently small, $T_s(X_0) \subset N_\delta(T_t(X_0))$ while $T_s(u) \subset N_\delta(T_t(u))$. On the other hand,

$$cl(T_s(X_0)) \subset N_\delta(T_s(X_0))$$

for any $\delta > 0$. Hence for such values of s , $C \subset Y - cl(T_s(X_0))$, while C contains points of the non-empty set $T_s(u)\backslash cl(T_s(X_0))$. Since T_s is full with respect to the pair (X,X_0) for each s , it follows that C is entirely contained in $T_s(X)$. Finally, we note that for s sufficiently near t , $y(s)$ lies in C . Hence $y(s)$ lies in $T_s(X)$ for s near t , and Q is open in $[0,1]$.

Since Q is both open and closed in $[0,1]$, $Q = [0,1]$. q.e.d.

Theorem 6. Let X be a set, X_0 a subset of X , T a mapping of X into 2^Y for a Banach space Y such that $T(X)$ is closed in Y . Suppose that $T(X\backslash X_0)$ is infinitesimally full in Y .

Then T is full with respect to the pair (X,X_0) .

Proof of Theorem 6. This is identical with the proof of Theorem 2 since the proof refers to the structure of the sets $T(X)$ and $T(X_0)$ in Y and not with the single-valued or multivalued character of the mappings of which these sets are the images.

The further results (Theorems 3 and 4) for single-valued mappings can be carried over to the multivalued case by analogous procedures.

References

[1] Browder, F.E., Normal solvability for nonlinear mappings in Banach spaces, Bull. Amer. Math. Soc., 77 (1971), 73-77.

[2] __________, Normal solvability and the Fredholm alternative for mappings into finite-dimensional manifolds, Jour. Funct. Anal., 8 (1971), 250-274.

[3] __________, Normal solvability and existence theorems for non-linear mappings in Banach spaces, Problems in Nonlinear Analysis, Ed. Cremonese, Roma, Italy, 1971, 19-35.

[4] __________, Normal solvability for nonlinear mappings and the geometry of Banach spaces, Problems in Nonlinear Analysis, Ed. Cremonese, Roma, Italy, 1971, 37-66.

[5] __________, Normal solvability and ϕ-accretive mappings of Banach spaces, Bull. Amer. Math. Soc., 78 (1972), 186-192.

[6] __________, Normally solvable nonlinear mappings in Banach spaces and their homotopies, Jour. Funct. Anal., 17 (1974), 441-446.

ON A THEOREM OF CARISTI AND KIRK

Felix E. Browder

In recent papers Caristi and Kirk have proved and applied the following theorem:

<u>Theorem</u>: Let M be a complete metric space with metric d, g a map of M into M (with g not necessarily continuous). Suppose that there exists a non-negative real-valued function ξ on M which is lower-semi-continuous such that for all x in M,

$$d(x,g(x)) \leq \xi(x) - \xi(g(x)) .$$

Then g has a fixed point in M.

Since they have given applications of this theorem to the study of inward mappings ([3], [4]) and of the theory of normal solvability ([1], [2], [5]) this result may well become an important tool in the future development of nonlinear functional analysis. On the other hand, their proof of the theorem is technical and complex being based upon an iterated use of transfinite induction.

It is our object in the present note to give a simple and transparent proof of the Caristi-Kirk Theorem which avoids the use of transfinite induction completely.

We begin by assuming that g has no fixed point on M. Then for each x in M, $d(x,g(x)) > 0$ implies that

$$\xi(g(x)) < \xi(x) .$$

For heuristic purposes, we note that the theorem has a trivial proof if g is continuous. In that case, if we set $x_j = g^j(x_0)$ for a fixed x_0 in M, we see that

$$d(x_{j+1},x_j) \leq \xi(x_j) - \xi(x_{j+1}) .$$

and $\xi(x_j)$ is decreasing in j. Summing from 0 to n, we see that

$$\sum_{j=0}^{n} d(x_j, x_{j+1}) \leq \xi(x_0) - \xi(x_{n+1}) \leq \xi(x_0) .$$

Hence,

$$\sum_{j=0}^{\infty} d(x_j, x_{j+1}) \leq \xi(x_0) ,$$

and $\{x_j\}$ is a Cauchy sequence in M. Since the metric is complete on M, x_j converges to a point x of M. Since $g(x_j) = x_{j+1}$, $g(x_j)$ also converges to x. If g is continuous, x must be a fixed point of g. In general, this need not be true, and the proof is obtained from a more abstract version of the iteration argument.

Proof of the Theorem: Suppose as before that g has no fixed points in M. We define an ordered set P whose elements are pairs (S,f) with S a countable subset of the interval $[0,\xi(x_0)]$ and f a mapping of S into M satisfying the following conditions:

(1) For each s in S, $\xi(f(s)) = s$. If $\{s_j\}$ is an increasing sequence in S converging to s, then s lies in S. The point $s_0 = \xi(x_0)$ lies in S and $f(s_0) = x_0$.

(2) For each s in S, if $x = f(s)$, then $s_1 = \xi(g(s))$ lies in S and $f(s_1) = g(s)$. Moreover, no point of the open interval (s_1, s) lies in S.

(3) If s_2 lies in S and s_2 is not the successor of some s in S as in condition (2), then $f(s_2) = \lim_{s>s_2, s\in S} f(x) = \lim_{s\to s_2^+} f(s)$ where $s_2^+ = \inf\{s\colon s\in S, s>s_2\}$.

(4) For any pair s_3, s_4 in S with $s_3 < s_4$,

$$d(f(s_3), f(s_4)) \leq s_4 - s_3 .$$

Since $\{S_0, f_0\}$ lies in P, where $S_0 = \cup\{\xi(g^j(x_0))\}$ and $f_0(\xi(g^j(x_0))) = g^j(x_0)$, P is non-empty.

We define the ordering on P by saying that (S,f) dominates (S_1, f_1) if S contains S_1 and $f_1 = f|S_1$.

Lemma 1. P is linearly ordered.

Proof of Lemma 1. Let (S,f) and (S_1,f_1) be two elements of P. Consider the family of half-open intervals $I = (s,\xi(x_0)]$ such that

$$S\cap I = S_1 \cap I \; ; \; f|_{S\cap I} = f_1|_{S_1\cap I} .$$

Such intervals exist since $(\xi(g(x_0)),\xi(x_0)]$ is such an interval. Since the union of the whole family of such intervals is itself an interval of the family, there is a largest such interval $(s,\xi(x_0)]$. If there exists no point s_1 of S_1 outside of I, it would follow that (S,f) dominates (S_1,f_1). Similarly, if there exists no point s_2 of S outside of I, then (S_1,f_1) dominates (S,f).

In the remaining case, let s_1 be the largest point of S_1 outside of I, s_2 the largest point of S outside of I. (Such largest points exist because of the closedness condition on S and S_1 in the condition (1) of the definition of elements of P.) By condition (3),

$$f_1(s_1) = \lim_{t>s_1,t\in S_1} f_1(t) = \lim_{t>s,t\in S_1} f_1(t)$$

$$= \lim_{t>s,t\in S} f(t) = \lim_{t>s_2,t\in S} f(t) = f(s_2) .$$

Hence

$$s_1 = \xi(f_1(s_1)) = \xi(f(s_2)) = s_2$$

and it follows immediately that S and S_1 coincide on the half-open interval $(\xi(g(f(s))),\xi(x_0)]$ and f and f_1 coincide there. This contradicts the assumption that I is the largest interval. q.e.d.

It follows from Lemma 1 that each element (S,f) of P is uniquely characterized by the first element s, and from the proof of Lemma 1, we see that S is uniquely determined by the smallest half-open interval $(s,\xi(x))]$ which contains it. Therefore, the family of elements of P may be written as a sequence $\{(S_j,f_j)\}$, and we may define

$$S' = \cup_j S_j ,$$

$$f': S' \to M , \quad f'|_{S_j} = f_j .$$

Since each S_j is countable, S' is countable. It follows by an obvious argument that the pair (S',f') satisfies the conditions (1), (2), (3), and (4) for elements of P. Hence (S',f') is an element of P and clearly the maximum element.

<u>Lemma 2</u>. Let $s' = \inf\{s: s\in S'\}$. Then:

(a) s' does not lie in S'.

(b) $\lim_{s>s',s\in S'} f'(s) = x'$ exists.

(c) If $s'' = \xi(x')$, then $s'' \le s'$. Setting $S'' = S' \cup \{\bigcup_{j=0}^{\infty} \xi(g^j(x'))\}$ and extending f' to a map f'' of S'' into M by setting $f''(s'') = x'$, we obtain an element (S'',f'') of P.

Proof of Lemma 2. Proof of (a): If s' lay in S', then $\xi(g(f'(s'))) < s'$ would also lie in S', which contradicts the definition of s'.

Proof of (b): By condition (4), for s_3, s_4 in S' with $s_3 < s_4$, we have

$$d(f'(s_4); f'(s_3)) \le s_4 - s_3 .$$

Hence $\{f'(s_j)\}$ is a Cauchy sequence for any s_j converging to s', and the limit is the element we denote by x'.

Proof of (c): The pair (S'',f'') is well-defined. We must show that it verifies the conditions (1), (2), (3) and (4) above for elements of P. The verification of conditions (1) and (2) is automatic. For condition (3), we need only verify the limit property for $s_2 = s''$ and then it follows from the definition of s'' and $f(s'')$. For the proof of condition (4), we note that for s_3, s_4 in S' with $s_3 < s_4$,

$$d(f(s_3), f(s_4)) \le s_4 - s_3 .$$

As s_3 tends to s', $f'(s_3) \to f''(s'')$ where $s'' \le s'$. Hence

$$d(f''(s''), f''(s_4)) \le s_4 - s' \le s_4 - s'' .$$

Hence condition (4) holds for the pair (S'',f'').

Proof of the Theorem completed: By Lemma 2, (S'',f'') is a proper extension of (S',f'), while the latter is maximal in P. This is a contradiction based upon the assumption that g has no fixed points. Hence g must have fixed points in M. q.e.d.

References

[1] Browder, F.E., Normal solvability for nonlinear mappings in Banach spaces, Bull. Amer. Math. Soc., 77 (1971), 73-77.

[2] Browder, F.E., Normal solvability and ϕ-accretive mappings of Banach spaces, Bull. Amer. Math. Soc., 78 (1972), 186-192.

[3] Caristi, J., Fixed theorems for mappings satisfying inwardness conditions, Trans. A.M.S., (to appear).

[4] Caristi,J. and W.A. Kirk, Geometric fixed point theory and inwardness conditions, Proc. Conf. on Geometry of Metric and Linear Spaces, Michigan State University, 1974, (to appear).

[5] Kirk,W.A. and J. Caristi, Mapping theorems in metric and Banach spaces, (to appear).

AN APPLICATION OF FIXED POINT THEORY ON A CERTAIN BIOMATHEMATICAL MODEL IN NONLINEAR PARTIAL DIFFERENTIAL EQUATIONS

Kuang-Ho Chen

Abstract

A system of equations which models nerve impulses is shown to have one and only one solution in the space $L_x^2(0,\infty) \times L_t^\infty(0,T)$ for every finite $T > 0$. However, the spatial norm of the solution, $L_x^2(0,\infty)$, is not necessarily bounded as T approaches infinity. The system is a mixed initial-boundary value problem obtained from the Hodgkin-Huxley model modified to include effects of capacitance and inductance. The methods of analysis include the Schauder fixed point theorem and the theory of semigroups in which the infinitesimal generator is time-dependent. The result for this system also gives existence and uniqueness of solutions to the modified scalar model.

1. Introduction

A system of two equations representing an abstract model of impulses in an unmyelinated squid axon is:

$$w_t = -Aw + C(w,x), \quad (x>0,\ t>0)$$

$$\equiv \begin{pmatrix} 0 & 1 \\ 1 & 0 \end{pmatrix} w_x + C(w,x); \tag{1.1}$$

$$w(x,0) = \Gamma(x), \quad (x > 0); \tag{1.2}$$

$$w(0,t) = G(t), \quad (t > 0). \tag{1.3}$$

Subsequently this mixed problem is referred to as problem (A). The function $C(w,x)$ is assumed to satisfy $c(0,x) = 0$ and the condition of Lipschitz-type

$$|C(w,x) - G(\bar{w},x)| \leq qe^{pt}|w(x,t) - \bar{w}(x,t)|\ K(t), \tag{1.4}$$

for all pairs w and $\bar{w}$, where $K(t)$ is a bounded, monotonically increasing function in t, $0 < t < t_o$, and is independent of w and $\bar{w}$. With $w = (w_1, w_2)^*$, the transpose of (w_1, w_2) the system represents a model derived from the Hodgkin-Huxley model [2], modified so as to include effects of core capacitance and inductance. This modified model is shown in the Appendix as equations (3.1) through (3.8), a quasi-linear wave equation coupled with three linear ordinary differential equations. The Appendix also shows a process of deriving the above system from the modified model and the relationships between the two problems. From (3.30) in the appendix, $K(t)$ is a polynomial of degree 1 in t. Let B_c be a Banach space with norm $||\cdot||_c$, which is either the space $L^\infty(0,c)$ or the space $C[0,c]$ in the maximum-norm. Denote by H_2 the product Hilbert space $L^2(0,\infty) \times L^2(0,\infty)$ and by $||\cdot||_2$ the related norm. Let H be the Banach space consisting of elements $X(\cdot,t) \varepsilon H_2$ for each t with norm

$$|||X||| = ||[||X(\cdot,t)||_2]||_c < \infty, \quad (0 < t < c = t_o) \tag{1.5}$$

for any finite t_o.

Theorem 1.1. In problem (A), let G belong to B_c, $(c = t_o > 0)$, with non-vanishing, differentiable second (or first) component; let F be in the space H_2; and assume F and G are compatible: $F(0) = G(0)$. Under assumption (1.4), problem (A) has exactly one solution w in H for each finite $t_o > 0$.

As a consequence of this Theorem, Theorem 3.1, and equation (3.18), the existence and uniqueness of solutions for the modified model (3.1) through (3.8) is also assured which is stated at the end of the Appendix as Theorem 3.3.

The next section is devoted to the proof of Theorem 1.1 as an application of the Schauder fixed point theorem and the theory of a semigroup of operators. Section 3 is the Appendix which illustrates the implications of Theorem 1.1 for the modified model.

2. An Application of the Schauder Fixed Point Theorem

The boundary condition in (1.3) will be adjusted so that the differential operator $A = -\begin{pmatrix} 0 & 1 \\ 1 & 0 \end{pmatrix} \frac{\partial}{\partial x}$ is skew selfadjoint in the Hilbert space H_2. Then, the initial value problem induced from (1.1) and (1.2) is written equivalently into a Volterra integral equation of the second kind, using the semigroup of operators with A as the infinitesimal generator. Finally, the fixed point theorem will provide existence and uniqueness of solutions.

Let $G(t) = (G_1, G_2)^*$. Then, the transformation given by

$$X = w - (G_1, 0)^* \tag{2.1}$$

and the operator $A(t) = -\begin{pmatrix} 0 & 1 \\ 1 & 0 \end{pmatrix} \frac{\partial}{\partial x}$, $(0 < t \leq t_o)$, with domain

$$D(A(t)) = \{X(\cdot,t) \in H_2 : \partial X(\cdot,t)/\partial x \in H_2 ; X(0,t) = (0, G_2(t))^*\}, \tag{2.2}$$

for any fixed differentiable $G_2(t) \neq 0$, carry problem (A) into the following initial value problem in H_2:

$$\frac{\partial X}{\partial t} + AX = P(X,x), \quad (t>0,\ x>0), \tag{2.3}$$

$$\equiv C(X-(G_1,0)^*, x); \tag{2.3'}$$

$$X(0) = \bar{F} \tag{2.4}$$

$$\equiv F-(G_1(0),0)^*. \tag{2.4'}$$

It is clear that for each fixed t, $A(t)$ is a skew selfadjoint operator, and $D(A(t))$ is a subspace of H.

Let $R(t)$ be the translation transformation in H_2 such that $R(t)X = X-(0,G_2)^*$. Using $R(t)$ instead of the Kato multiplication transformation [3; Theorem 2, p. 483], we conclude the following theorem.

Theorem 2.1. Under assumptions of Theorem 1.1 on F and G, there exists a unique operator valued function $U(t,s)$ over H_2 and defined in the triangle $0 \leq s \leq t \leq t_o$, with the following properties: (a) $U(t,s)$ is a bounded linear operator and is strongly continuous in (t,s) and $U(t,t) = I$;

(b) $U(t,r) = U(t,s)U(s,r)$, $(0 \leq r \leq s \leq t \leq t_o)$;

(c) $U(t,s)$ maps $D(A(s))$ into $D(A(t))$, and for each X_o in $D(A(s))$: $X(t) = U(t,s)X_o$ is strongly differentiable for $t > s$; $\partial X/\partial t = -A(t)X$; and $A(t)X(t)$ is strongly continuous in t.

Consequently, the initial value problem (2.3) and (2.4) is equivalent to the following integral equation

$$X(t) = U(t,0)\bar{F} + \int_0^t U(t,s)P(X(s),x)dx, \quad (0 \leq t \leq t_o). \tag{2.5}$$

For fixed $\bar{F}$ and finite $t_o > 0$, define the integral transformation in H as follows:

$$T[X](x,t) = U(t,0)\bar{F}(x) + \int_0^t U(t,s)P(X(x,s),s)ds \;. \tag{2.6}$$

For each pair X and Y in $D(A(t))$, $(0 \le t \le t_o)$, the Minkowski inequality for integrals, Theorem 2.1, and then condition (1.4) yield

$$\begin{aligned} ||T[X](\cdot,t)-T[Y](\cdot,t)||_2 &\le C \int_0^t ||P(X(\cdot,s),\cdot)-P(Y(\cdot,s),\cdot)||_2 ds \\ &\le C\ q \int_0^t e^{ps} K(s)\,||X(\cdot,s) - Y(\cdot,s)||_2 ds \\ &= \int_0^t K_1(s)\,||X(\cdot,s) - Y(\cdot,s)||_2 ds \;; \end{aligned} \tag{2.7}$$

where the constant C depends only on $A(t)$ and

$$K_1(s) = C\ q\ e^{ps}\ K(s)\ ,\ 0 \le s \le t_o \;. \tag{2.8}$$

From (2.7) with $Y = 0$, $T[X](\cdot,t) \in H_2$. This and (2.6) imply $T[X](\cdot,t) \in D(A(t))$. Applying (2.7) twice gives

$$||T^2[X](\cdot,t) - T^2[Y](\cdot,t)||_2 \le \int_0^t K_2(t,r)\,||X(\cdot,r) - Y(\cdot,r)||_2 dr, \tag{2.9}$$

where

$$K_2(t,r) = K_1(r) \int_r^t K_1(s)ds,\ (0 \le r < t \le t_o) \;. \tag{2.10}$$

Similar argument shows $T^2[X](\cdot,t) \in D(A(t))$ if $X(\cdot,t) \in D(A(t))$, $(0 \le t \le t_o)$. Inductively, we find

$$\begin{aligned} &||T^{n+1}[X](\cdot,t) - T^{n+1}[Y](\cdot,t)||_2 \\ &\qquad \le \int_0^t K_{n+1}(t,r)\,||X(\cdot,r) - Y(\cdot,r)||_2 dr \;, \end{aligned} \tag{2.11}$$

where

$$K_{n+1}(t,r) = K_1(r) \int_r^t K_n(t,s)ds\ ,\ (0 \le r \le t \le t_o)\ , \tag{2.12}$$

and $T^n[X](\cdot,t)$ belongs to $D(A(t))$, $(0 \le t \le t_o)$. By (2.8), $K_1(t) \le C|q|\ \exp\{pt_o\}K(t_o)$, $(0 \le t \le t_o)$. Similarly, by (2.10), $K_2(t,r) \le [C|q|\ \exp\{pt_o\}K(t_o)]^2(t-r)$, and inductive employment of (2.12) leads to the estimate

$$K_{n+1}(t,r) \le [C|q|\ \exp\{pt_o\}K(t_o)]^{n+1}(t-r)^n/n! \;. \tag{2.13}$$

Hence, for sufficiently large n and for $0 \le r < t \le t_o$ with finite fixed $t_o > 0$,

$$|||T^n[X] - T^n[Y]||| \le C_n|||X - Y||| \;; \tag{2.14}$$

$$C_n = [C|q| \exp\{pt_o\}K(t_o)t_o]^n/n! < 1 \ . \tag{2.15}$$

Therefore, for large n depending on t_o, T^n is a contraction transformation on $D(A(t))$, $(0 \leq t \leq t_o)$. Consequently, we have the following assertion which implies Theorem 1.1.

Theorem 2.2. The transformation T defined by (2.6) has a fixed point in H for each F in H_2 and G in B_c, $(c = t_o > 0)$ such that F and G are compatible at $t = x = 0$ and $G_2 \neq 0$ and G_2 is differentiable on $[0,t_o]$.

Since C_n grows with t_o exponentially, the $L_x^2(0,\infty)$-norm of the solution may not be bounded as $t \to \infty$.

3. Appendix

In 1967, H.M. Lieberstein [4] modified the equations (cf. Hodgkin-Huxley [2]) for impulses in an unmyelinated squid axon to include effects of core capacitance and inductance without introducing any new parameter. The modified model is a mixed initial-boundary value problem, a quasi-linear wave equation coupled with three linear ordinary differential equations:

$$v_{xx} = p[v_t + qB(v,x)] + [v_t + qB(v,x)]_t, \quad (x > 0, \ t > 0) \ , \tag{3.1}$$

$$v(x,0) = f_1(x), \ f_t(x,0) = f_2(x), \ (x > 0) \ , \tag{3.2}$$

$$v(0,t) = g_1(t), \ v_x(0,t) = g_2(t), \ (t > 0) \ , \tag{3.3}$$

where the non-linear term $B(v,x)$ is given by

$$B(v,x) = q_1 n_1^4 (v-\bar{v}_1) + q_2 n_2^3 n_3 (v-\bar{v}_2) + q_3 (v-\bar{v}_3) \tag{3.4}$$

with q_i and $\bar{v}_1$ as six given constants; the n_i, (i=1,2,3) have values in [0,1] and satisfy the equations

$$\frac{dn_i}{dt} = \alpha_i(1-n_i) - \beta_i n_i \ , \ (t > 0, \ i=1,2,3) \ ; \tag{3.5$_i$}$$

$$n_i(0) = \Phi_i(f_1) \ , \tag{3.6$_i$}$$

and the six positive functions of v: α_i and β_i, are given as follows:

$$\alpha_i(v) = a_i[0.1v+b_i][-1+\exp\{0.1v+b_i\}]^{-1}, \ (i = 1,2) \ , \tag{3.7$_i$}$$

$$\beta_i(v) = c_i \exp\{d_i v\} \ , \ (i = 1,2) \ , \tag{3.8$_i$}$$

$$\alpha_3(v) = 0.07 \exp\{v/20\} , \tag{3.7$_3$}$$

$$\beta_3(v) = [1 + \exp\{0.1v + 3\}]^{-1} , \tag{3.8$_3$}$$

with $a_1 = 0.1$, $b_1 = 1$, $c_1 = 0.125$, $d_1 = 0.0125$ and $a_2 = 1$, $b_2 = 2.5$, $c_2 = 4$, $d_2 = 1/18$. Let $n(t)$ represent the three functions $n_i(t)$ whenever there is no ambiguity. It is clear from (3.2) and $(3.5)_i$ that, for each differentiable function in $L^2(0,\infty) \times B_c$, $(c = t_o > 0)$,

$$n(t) = \Phi(f_1) \exp\{-\int_o^t \alpha(s) + \beta(s)ds\} + \int_0^t \alpha(s) \exp\{-\int_s^t \alpha(r) + \beta(r)dr\}ds , \quad (t > 0), \tag{3.9}$$

which is analytic in v because α's and β's are analytic in v .

The above scalar model is transformed as follows into matrix form. Define

$$u(x,t) = \int_0^t e^{ps} v(x,s)ds + p \int_0^x (x-y) f_1(y)dy \tag{3.10}$$

and define the vector

$$w(x,t) = (u_t(x,t) , u_x(x,t))^* . \tag{3.11}$$

Then, with $A = -\begin{pmatrix} 0 & 1 \\ 1 & 0 \end{pmatrix} \partial/\partial x$,

$$w_t + Aw = C(w,x;f_1,f_2) , \quad (x > 0 , t > 0)$$
$$\equiv (-qe^{pt}B(e^{-pt}w_1,x) - p(w_1-f_1) - qB(f_1,x) - f_2, 0)^* , \tag{3.12}$$

$$w(x,0) = F(x) \equiv (f_1(x), p\int_0^x f_1(y)dy)^* , \quad (x > 0) , \tag{3.13}$$

$$w(0,t) = G(t) \equiv (e^{pt}g_1(t), \int_0^t e^{ps}g_2(s)ds)^*, \quad (t > 0) . \tag{3.14}$$

These are just system (1.1), (1.2), and (1.3), which is problem (A).

In section 2, $A = A(t)$ is required to be skew selfadjoint with either of the following two as domain in the Hilbert space H_2 :

$$D_1(A) = \{X \in H_2 \colon X_x \in H_2, X_1(0) = 0\} ; \tag{3.15$_1$}$$

$$D_2(A) = \{X \in H_2 \colon X_x \in H_2, X_2(0) = 0\} . \tag{3.15$_2$}$$

Therefore, the respective translations are defined by

$$X(x,t) = w(x,t) - (G_1(t),0)^* ; \tag{3.16$_1$}$$

$$X(x,t) = w(x,t) - (0,G_2(t))^* . \tag{3.16$_2$}$$

In terms of this translation operator, problem (A) has the representation as an initial value problem in the form of (2.3) and (2.4):

$$X_t + AX = P(X,x) \ , \ (x > 0) \tag{3.17}$$

$$\equiv \begin{cases} C(X+(G_1,0)^*, \ x; \ f_1,f_2)-(0,G_1')^* \ ; \\ C(X+(0,G_2)^*, \ x; \ f_1,f_2)-(G_2,0)^* \ , \end{cases} \tag{3.17'}$$

$$X(x,0) = \bar{F}(x) \equiv F(x)-(G_1(0),0)^* \quad \text{or} \quad F(x)-(0,G_2(0))^* \ . \tag{3.18}$$

Problem (A) and this representation are related to the scalar model by the following:

Theorem 3.1. (a) If v is a solution of system (3.1), (3.2), and (3.3), then w defined by (3.11) with u given by (3.10), is a solution of system (3.12), (3.13) and (3.14).
(b) If w is a solution of (3.12), (3.13), and (3.14) and if w_{1x} and w_{2t} are continuous and $w_{2tx} = w_{2xt}$, then system (3.1), (3.2), and (3.3) has a solution v such that the u given by (3.10) satisfies relation (3.11).
(c) System (3.1), (3.2), and (3.3) is equivalent to system (3.17) and (3.18) with domain of A defined by $(3.15)_i$ and $(3.16)_j$, $(j=1,2)$.

Proof. Assertions (a) and (c) can be proved by routine steps together with the above derivations of the new representations.

For assertion (b), let u be the function $u(x,t) = \int_0^x w_2(y,t)+C(t)$ for some differentiable function $C(t)$ to be determined. Since $w_{2t}(y,t)$ is continuous, $u_t(x,t)$ exists and satisfies the relation

$$u_t(x,t) = \int_0^x w_{1y}(y,t)\,dy + C'(t) \ .$$

In this, $w_{1x} = w_{2t}$ is used. Putting $C'(t) = e^{pt}g_1(t)$, then $C(t) = \int_0^t e^{pt}g_1(s)\,ds$, and $u(x,t)$ becomes

$$u(x,t) = \int_0^x w_2(y,t)\,dy + \int_0^t e^{ps}g_1(s)\,ds \ .$$

Set the function v to be

$$\begin{aligned} v(x,t) &= e^{-pt}u_t(x,t) \\ &= e^{-pt}\int_0^x w_{2t}(y,t)\,dy + g_1 \\ &= e^{-pt}[w_1(x,t)-w_1(0,t)]+g_1(t) = e^{-pt}w_1(x,t) \ . \end{aligned} \tag{3 19}$$

In this, $w_{1x} = w_{2t}$ and (3.14) are used. It remains to show that v satisfies system (3.1), (3.2), and (3.3). By (3.14),

$$v(0,t) = g_1(t); \ v(x,0) = w_1(x,0) = f_1(x) \ . \tag{3.20}$$

These two relations are the first relations of (3.3) and (3.2), respectively. By (3.19) and the second component of (3.11),

$$v_t(x,t) = [-pw_1(x,t) + w_{1t}(x,t)]e^{-pt}$$
$$= e^{-pt}[w_{2x}(x,t) - pf_1(x) + f_2(x)]$$
$$- q[B(v(x,t),x) - B(f_1(x),x)e^{-pt}] . \tag{3.21}$$

Taking $t = 0$ and using (3.14) and (3.20) give

$$v_t(x,0) = w_{2x}(x,0)-pf_1(x)+f_2(x)-q[B(v(x,0),x)-B(f_1(x),x)]$$
$$= f_2(x)$$

which is the second condition in (3.2). By (3.19),

$$v_x(x,t) = e^{-pt}\ w_{1x}(x,t) = e^{-pt}\ w_{2t}(x,t) . \tag{3.22}$$

Setting $t = 0$ and using (3.13) yield

$$v_x(0,t) = e^{-pt}\ w_{2t}(0,t) = g_2(t) ,$$

which is the second relation of (3.3). For (3.1), derivatives of (3.22) imply

$$v_{xx} = e^{-pt}\ w_{2tx} ;$$
$$[v_t+qB(v,x)]_t = \{e^{-pt}[w_{2x}-pf_1+f_2+qB(f_1,x)]\}_t$$
$$= e^{-pt}[w_{2xt}-p\ w_{2x}-pf_1+f_2+qB(f_1,x)\ ] .$$

These results, (3.22), and the condition $w_{2tx} = w_{2xt}$ lead to

$$p[v_t+qB(v,x)]+[v_t+qB(v,x)]_t = e^{-pt}w_{2xt} = v_{xx} ,$$

which is (3.1), and the proof is complete.

It remains to verify that the Lipschitz-type condition for $P(X,x)$ satisfies assumption (1.4). By (3.17) and (3.11),

$$|P(X,x)-P(y,x)| = |C(X+F_1,x,t;f_1,f_2)-C(Y+F_1,x,t;f_1f_2)|$$
$$= |q|e^{pt}|B(v_1(x,t),x)-B(v_2(x,t),x)| . \tag{3.23}$$

Therefore, (3.4) and $0 \le n_i \le 1$ imply

$$|P(X,x)-P(Y,x)| \le |q|e^{pt}|v_1(x,t)-v_2(x,t)|[|q_3|+|q_1|\{1+4\ n_1'(z)|(|\bar{v}_1|+|z|)\}$$
$$+ |q_2|\{1+(3|n_2'(z)|+|n_3'(z)|)(|\bar{v}_2|+|z|)\}] , \tag{3.24}$$

where $z = \theta v_1 + (1-\theta)v_2$, with $0 \le \theta \le 1$, and $n_i'(z)$ denotes the partial derivative of n_i with respective to v and evaluated at z. By $0 \le \Phi(f_1) \le 1$ in (3.9), with notation $\alpha(s) = \alpha(v(s))$ and

$\beta(s) = \beta(v(s))$,

$$|n'(t)| \leq \int_0^t |\alpha'(s)| + |\beta'(s)|ds \exp\{-\int_0^t \alpha(s)ds\} \tag{3.25}$$
$$+ \int_0^t [|\alpha'(s)|+\alpha(s)\int_s^t |\alpha'(r)|+|\beta'(r)|dr]\exp\{-\int_s^t \alpha(r)+\beta(r)dr\}ds \ .$$

Estimates for the right hand side of (3.24) are now derived. Consider $i = 1,2$ first. By $(3.7)_i$ and $(3.8)_i$,

$$\alpha_i'(v) = -0.1a_i[1+(b_i-1+0.1v)\exp\{0.1v+b_i\}][-1+\exp\{0.1v+b_i\}]^{-2} \ ; \tag{3.26}$$

$$\beta_i'(v) = c_i d_i \exp\{d_i v\} \ . \tag{3.27}$$

By the usual graphing process in calculus, $(3.7)_i$ and $(3.25)_i$ imply $-0.1 < \alpha_i'(z)/\alpha_i(z) < 0$, and this estimate is sharp. Hence for both exponents $h = 0$ and 1 , this estimate with those in $(3.25)_i$ and $(3.26)_i$ gives

$$|v^h(t)n_i'(v(t))| \leq |v(t)|^h \exp\{\frac{-1}{2}\int_0^t \alpha(s) + \beta(s)ds\} \ .$$
$$\cdot([\int_0^t 0.1\alpha(s)ds[\exp\{\frac{-1}{2}\int_0^t \alpha(s)ds\} + [\int_0^t d_i\beta(s)ds]\exp\{\frac{-1}{2}\int_0^t \beta(s)ds\})$$
$$+ \int_0^t [\alpha(s)\exp\{\frac{-1}{3}\int_s^t \alpha(r)dr\}][|v(t)|^h\exp\{\frac{-1}{3}\int_s^t \alpha(r) + \beta(r)dr\}] \ .$$
$$\cdot[0.1 + \int_s^t 0.1\ \alpha(r)dr\ \exp\{\frac{-1}{3}\int_s^t \alpha(r)dr\}+\int_s^t d_i\beta(r)dr\ \exp\{\frac{-1}{3}\int_s^t \beta(r)dr\}]ds$$
$$\leq (0.4+3d_i)(2|v(t)|^h\{\int_0^t \alpha(s) + \beta(s)ds\}^{-1}$$
$$+ 9\int_0^t [\alpha(s)\{\int_s^t \alpha(r)dr\}^{-1}][|v(t)|^h\{\int_s^t \alpha(r)+\beta(r)dr\}^{-1}]ds \ .$$

For $h = 0$, the first and third of the expressions $\{\ \}^{-1}$ have an upper bound of unity. For $h = 1$, because α and β are analytic in v and v is a smooth function in t , in the first and third $\{\ \}$ the integrals are dominated above by the integral over a small interval near t and $v(s)$, $v(r)$ possess values near $v(t)$. Moreover, an appropriate bound for $\beta(s)$ and $\beta(r)$ is 0 if $v(t) > 0$ and is also 0 for $\alpha(s)$ and $\alpha(r)$ if $v(t) > 0$. On the other hand, in the first bracket the integral is dominated below by the one with small interval around 0 and $\alpha(s)$ deviates slightly from $\alpha(0)$. Therefore, for all v ,

$$|v^h(t)n_i'(v(t))| \leq k_i(t) \quad , \quad (h = 0,1; \ i = 1,2) \ ; \tag{3.28$_i$}$$

$$k_i(t) = (r + 30d_i)(1 + t) \quad , \ (i = 1,2) \ . \tag{3.29$_i$}$$

Now consider the case i=3. From $(3.7)_3$, $\alpha_3'(v)=0.07\exp\{v/20\}/20=0.05\alpha_3(v)$ and from $(3.8)_3$, $|\beta_3'(v)| = |-0.1\{\exp\{0.1v+3\}[1+\exp\{0.1v+3\}]^{-2} \leq 0.1\beta_3(v)$. By an argument similar to that for $(3.28)_i$, these two relations and (3.25) lead to the estimate,

$$|v^h(t)n_3'(v(t))| \leq 2.7|v(t)|^h\{\int_0^t \alpha(s) + \beta(s)ds\}^{-1}$$
$$+ 4.5\int_0^t \alpha(s)\{\int_s^t \alpha(r)dr\}^{-1}|v(t)|^h\{\int_s^t \alpha(r)+\beta(r)dr\}^{-1}ds$$
$$\leq 5(1+t) \ , \ (h = 0,1) \ . \tag{3.28$_3$}$$

$$k_3(t) = 5(1+t) \ . \tag{3.29$_3$}$$

Therefore, (3.24) and $(3.28)_i$ yield

$$|P(X,x) - P(Y,x)| \leq |q|e^{pt}|v_1(x,t) - v_2(x,t)|[|q_3| +$$
$$+ |q_1|\{1+4k_1(t)(|\bar{v}_1|+1\}+|q_2|\{1+(3k_2(t)+k_3(t))(|\bar{v}_2|+1)\}] \ ,$$

and we conclude the following statement:

Theorem 3.2. There is a monotonically increasing positive function $K(t)$ dependent only on physiological constants $\bar{v}'$ and q's such that

$$|P(X,x)-P(Y,x)| \leq |q|e^{pt}K(t)|X(x,t)-Y(x,t)| \ ; \tag{3.30}$$

$$K(t) = |q_3|+|q_2|\{1+(3k_2(t)+k_3(t))(|\bar{v}_2|+1)\}+|q_1|\{1+4k_1(t)(|\bar{v}_1|+1)\}. \tag{3.31}$$

As a consequence of Theorems 1.1, 3.1, and 3.2, we state the existence and uniqueness of solutions for system (3.1) through (3.8).

Theorem 3.3. Assume that functions f_i in (3.2) are $C^{(1)}[0,\infty)$ and satisfy the condition

$$(f_1(x), \ p\int_0^x f_1(y)dy)^* \ \varepsilon \ H_2 \ ; \ f_2 \ \varepsilon \ L^2(0,\infty) \ , \tag{3.31}$$

and that functions g_i in (3.3) are $C^{(1)}[0,T]$ and satisfy the condition

$$(e^{pt}g_1(t), \int_0^t e^{ps}g_2(s)ds)^* \ \varepsilon \ B_T \ , \ (0 < t < T) \ . \tag{3.32}$$

Then, system (3.1) through (3.8) has exactly one solution belonging to

$L^2(0,\infty) \times B_T$, $(T > 0)$.

Proof: Since f_i and g_i are $C^{(1)}$, it is clear from relations (3.12)', (3.13), and (3.14) and then from (3.16), (3.17)', and (3.18) that the solution to Problem (A) is smooth.

Particularly, two mixed derivatives of the second component of this solution of Problem (A) are smooth and then are equal. Therefore, the conditions in (b) of Theorem 3.1 are fulfilled. Thus, system (3.1) through (3.8) has exactly one solution. Here, the smoothness of v in t supports the solution of (3.5) with the representation (3.9) through Theorem 3.2. The properties of the solution w is carried through by relations (3.19), (3.13), and (3.14). Then, the proof of the theorem is complete.

It is important to remark that this paper is essentially in mathematics, not in physiology, because the modified model (3.1) through (3.8) has been justified only theoretically and numerically by computer, and not by laboratory measurements. On the other hand, the Hodgkin-Huxley model [2] has substantial experimental support. The existence and uniqueness of solutions of this model has been studied by Chen [1].

4. Bibliography

[1] Chen, K.H., Existence and uniqueness theorems of the Hodgkin-Huxley model on the propagation of nerve impulses (to appear).

[2] Hodgkin, A.L. and Huxley, A.F., A quantitative description of membrane current and its application to conduction and excitation in nerve, J. Physiol. 117 (1952), 500-544.

[3] Kato, T., On linear differential equations in Banach spaces, Comm. Pure Appl. Math. 9 (1956), 479-486.

[4] Lieberstein, H.M., On the Hodgkin-Huxley partial differential equation, Math. Biosciences, 1 (1967), 45-69.

Department of Mathematics
University of New Orleans
New Orleans, Louisiana 70122

REMETRIZATION AND FAMILY OF COMMUTING CONTRACTIVE TYPE MAPPINGS

Kim-Peu Chew and Kok-Keong Tan[1]

Abstract

Let X be a compact metrizable space and $\{f_n: n=1,2,...\}$ be a commuting family of continuous f-nctions on X. Suppose $\{f_n: n=1,2,...\}$ has a common fixed point x_0 in X such that $\limsup_{n\to\infty} f_n[X] = \{x_0\}$ and $\bigcap_{n=1}^{\infty} f_k^n[X] = \{x_0\}$ for all $k = 1,2,...$. Then for each $\alpha \in (0,1)$, there exists a metric d on X inducing the same topology on X such that

$$d(f_n(x),f_n(y)) \leq \alpha d(x,y), \text{ for all } x,y \in X \text{ and } n=1,2,... .$$

The above result can also be extended to a countably compact Tychonov space.

Throughout this note, (X,ζ) will denote a Tychonov space and $\mathcal{P}(\zeta)$ will denote the collection of all families of pseudometrics on X generating the topology ζ on X. Let $f: X \to X$. Then f is said to be (a) squeezing [2] iff $\bigcap_{n=1}^{\infty} f^n[X]$ is a singleton, and (b) topologically squeezing [3] iff $\bigcap_{n=1}^{\infty} \overline{f^n[X]}$ is a singleton, where for $M \subseteq X$, $\bar{M}$ denotes the ζ-closure of M. In [2], it was proved that for a compact space (X,ζ), a continuous mapping $f: X \to X$ is squeezing iff f is a topological c-contraction for any $c \in (0,1)$, i.e., for any $c \in (0,1)$, there exists $D \in \mathcal{P}(\zeta)$ such that $\rho(f(x),f(y)) \leq c\rho(x,y)$ for all $\rho \in D$ and all x,y in X. This result has been extended in [3] by weakening the condition "compact Hausdorff" to "countably compact Tychonov" and replacing "squeezing" by "topologically squeezing." It is the purpose of this note to extend the result in [3] to a countable family of selfmaps on X instead of a single map.

Theorem 1. Let (X,ζ) be a countably compact space and let $\{f_n | n = 1,2,\ldots\}$ be a commuting family of continuous functions on X with a common fixed point x_0 in X such that

(i) $\bigcap_{n=1}^{\infty} \overline{f_k^n[X]} = \{x_0\}$ for each $k = 1,2,3,\ldots$

and (ii) $\limsup_{n\to\infty} f_n[X] = \{x_0\}$,

then for any $D \in P(\zeta)$ there exists $D^* \in P(\zeta)$ such that Card D = Card D^* and F is nonexpansive with respect to D^*, i.e., $\rho(f_n(x),f_n(y)) \leq \rho(x,y)$ for all $\rho \in D^*$, $n=1,2,3,\ldots$ and for all x,y in X .

Proof: Let $D \in P(\zeta)$. We may assume that $d \leq 1$ for each $d \in D$, otherwise we replace d by the equivalent pseudometric $d/(1+d)$. For $d \in D$, we define

$$d^*(x,y) = \sup\{d(f_1^{k_1}\ldots f_n^{k_n}(x), f_1^{k_1}\ldots f_n^{k_n}(y)) \,|\, k_i \geq 0,\ n=1,2,3,\ldots\}$$

for all x,y in X . Clearly, $d(x,y) \leq d^*(x,y)$ and $d^*(f_n(x),f_n(y)) \leq d^*(x,y)$ for all x,y in X and $n = 1,2,3,\ldots$. We shall show that $D^* = \{d^* | d\in D\} \in P(\zeta)$. To do this, it is sufficient to show that for any convergent net $\{x_\alpha\} \to x$ $(\alpha\in\Lambda)$ the following proposition is true:

(*) $\forall\, d^* \in D^*$, $\forall\, \varepsilon > 0$, $\exists\, \alpha \in \Lambda$ such that $d^*(x_\beta,x) < \varepsilon$ for all $\beta > \alpha$, where $>$ denotes the partial order in Λ .

If this were not true, there would exist $d^* \in D^*$ and $\varepsilon > 0$ such that for each $\alpha \in \Lambda$, $d^*(x_{\beta(\alpha)},x) \geq \varepsilon$ for some $\beta(\alpha) > \alpha$. This would mean that for each $\alpha \in \Lambda$, there would exist nonnegative integers $N(\alpha)$, $k_1(\alpha)$, $k_2(\alpha),\ldots,k_{N(\alpha)}(\alpha)$ with $k_{N(\alpha)}(\alpha) \neq 0$ such that

$$(**) \quad d(f_1^{k_1(\alpha)}\ldots f_{N(\alpha)}^{k_{N(\alpha)}(\alpha)}(x_{\beta(\alpha)}),\ f_1^{k_1(\alpha)}\ldots f_{N(\alpha)}^{k_{N(\alpha)}(\alpha)}(x))$$

$$> d^*(x_{\beta(\alpha)},x) - \frac{\varepsilon}{2} \geq \frac{\varepsilon}{2} .$$

Now, we shall distinguish two cases.

Case 1. Suppose that $\{N(\alpha) | \alpha\in\Lambda\}$ is unbounded. Then there exists $\alpha_i \in \Lambda$ with $N(\alpha_1) < N(\alpha_2) < N(\alpha_3) < \ldots$ and we see that both points

$f_1^{k_1(\alpha_i)} \cdots f_{N(\alpha_i)}^{k_{N(\alpha_i)}(\alpha_i)}(x_{\beta(\alpha_i)})$ and $f_1^{k_1(\alpha_i)} \cdots f_{N(\alpha_i)}^{k_{N(\alpha_i)}(\alpha_i)}(x)$ belong to $f_{N(\alpha_i)}[X]$ for each $i = 1,2,3,\ldots$. Since X is countably compact, these two sequences $(f_1^{k_1(\alpha_i)} \cdots f_{N(\alpha_i)}^{k_{N(\alpha_i)}(\alpha_i)}(x_{\beta(\alpha_i)}))_{i=1}^{\infty}$, $(f_1^{k_1(\alpha_i)} \cdots f_{N(\alpha_i)}^{k_{N(\alpha_i)}(\alpha_i)}(x))_{i=1}^{\infty}$ and any of their subsequences have cluster points which must be in $\limsup_{n\to\infty} f_n[X] = \{x_0\}$. This contradicts (**).

Case 2. Suppose that $\{N(\alpha)\,|\,\alpha\in\Lambda\}$ is bounded. Let $M = \max\{N(\alpha)\,|\,\alpha\in\Lambda\}$. Then from (**), we have:

$$(\dagger) \qquad d(f_1^{k_1(\alpha)} \cdots\cdot f_M^{k_M(\alpha)}(x_{\beta(\alpha)}), f_1^{k_1(\alpha)} \cdots f_M^{k_M(\alpha)}(x)) \geq \frac{\varepsilon}{2} \ \forall \ \alpha\in\Lambda,$$

where $k_i(\alpha) \geq 0 \ \forall \ i=1,2,\ldots,M$ and $\alpha\in\Lambda$; (for $N(\alpha) < \ell \leq M$, let $k_\ell(\alpha) = 0$) .

Suppose there exists $i \in \{1,2,\ldots,M\}$ such that $\{k_i(\alpha)\,|\,\alpha\in\Lambda\}$ is unbounded. Then $1 \leq k_i(\alpha_1) < k_i(\alpha_2) < k_i(\alpha_3) < \ldots$ for some $\alpha_j \in \Lambda$, $j=1,2,3,\ldots$. We see that both $f_1^{k_1(\alpha_j)} \cdots f_M^{k_M(\alpha_j)}(x_{\beta(\alpha_j)})$ and $f_1^{k_1(\alpha_j)} \cdots f_M^{k_M(\alpha_j)}(x)$ belong to $f_i^{k_i(\alpha_j)}[X]$ for each $j = 1,2,\ldots,$ and since $\bigcap_{j=1}^{\infty} \overline{f_i^{k_i(\alpha_j)}[X]} = \bigcap_{n=1}^{\infty} \overline{f_i^n[X]} = \{x_0\}$ these two sequences $(f_1^{k_1(\alpha_j)} \cdots f_M^{k_M(\alpha_j)}(x_{\beta(\alpha_j)}))_{j=1}^{\infty}$, $(f_1^{k_1(\alpha_j)} \cdots f_M^{k_M(\alpha_j)}(x))_{j=1}^{\infty}$ and any of their subsequences must have the same cluster point x_0 . This contradicts (†). Therefore, for each $i=1,2,\ldots,M$, the set $\{k_i(\alpha)\,|\,\alpha\in\Lambda\}$ is bounded, say $\sup\{k_i(\alpha) : \alpha\in\Lambda\} = N_i$, for each $i=1,\ldots,M$. Thus there exists a cofinal subset Λ_0 of Λ and a positive integer n_1 such that $k_1(\alpha) = n_1$ for all $\alpha\in\Lambda_0$. We note that $\{x_{\beta(\alpha)}\,|\,\alpha\in\Lambda_0\}\to x$. Thus for each $j_i \leq N_i$, $i=2,\ldots,M$,

$$f_1^{n_1} f_2^{j_2} \cdots f_M^{j_M}(x_{\beta(\alpha)}) \to f_1^{n_1} f_2^{j_2} \cdots f_M^{j_M}(x) \qquad (\alpha\in\Lambda_0)$$

so that there exists $\alpha_0(j_2,\ldots,j_M) \in \Lambda_0$ such that

$$d(f_1^{n_1} f_2^{j_2} \cdots f_M^{j_M}(x_{\beta(\alpha)}) \ , \ f_1^{n_1} f_2^{j_2} \cdots f_M^{j_M}(x)) < \frac{\varepsilon}{2}$$

for all $\alpha \in \Lambda_0$ with $\alpha_0(j_2,\ldots,j_M) < \alpha$. Choose any $\alpha_0 \in A_0$ with $\alpha_0(j_2,\ldots,j_M) < \alpha_0$ for all $j_i \le N_i$ and $i=2,\ldots,M$. It follows that for all $\alpha \in \Lambda_0$ with $\alpha_0 < \alpha$

$$d(f_1^{k_1(\alpha)} \cdots f_M^{k_M(\alpha)}(x_{\beta(\alpha)}) , f_1^{k_1(\alpha)} \cdots f_M^{k_M(\alpha)}(x)) < \frac{\varepsilon}{2}$$

which again contradicts (†).

This completes the proof that $D^* \in P(\zeta)$.

Corollary 2. Let (X,ζ) be a compact metrizable space and $\{f_n: n=1,2,\ldots\}$ be a commuting family of continuous functions on X with a common fixed point x_0 in X such that

(i) $\bigcap_{n=1}^{\infty} f_k^n[X] = \{x_0\}$ for each $k=1,2,\ldots$

and (ii) $\limsup_{n\to\infty} f_n[X] = \{x_0\}$

then there exists a metric d on X generating the topology ζ on X such that $d(f_n(x),f_n(y)) \le d(x,y)$ for all x,y in X and $n=1,2,3,\ldots$.

Theorem 3. Let (X,ζ) be a compact metrizable space and $\{f_n: n=1,2,\ldots\}$ be a commuting family of continuous functions on X with a common fixed point x_0 in X such that

(i) $\bigcap_{n=1}^{\infty} f_k^n[X] = \{x_0\}$ for each $k=1,2,3,\ldots$

and (ii) $\limsup_{n\to\infty} f_n[X] =\{x_0\}$

then for any $\alpha \in (0,1)$ there exists a metric d on X generating the topology ζ on X such that $d(f_n(x),f_n(y)) \le \alpha d(x,y)$ for all x,y in X and $n=1,2,3,\ldots$.

Proof: By the above Corollary, there is a metric ρ on X generating the topology ζ on X such that $\rho(f_n(x),f_n(y)) \le \rho(x,y)$ for all x,y in X and $n=1,2,3,\ldots$. For $x,y \in X$, let

$n(x) = \max\{m: m = n_1+n_2+\ldots+n_k$ where $n_i \ge 0$, $k \ge 1$ and $x \in f_1^{n_1}\ldots f_k^{n_k}[X]\}$

$n(x,y) = \min(n(x), n(y))$

$\lambda(x,y) = \alpha^{n(x,y)}\rho(x,y)$ (with the convention that $\alpha^{\infty} = 0$) .

Note that: (1) $n(x)$ is finite for all $x \ne x_0$;

(2) $n(f_m(x),f_m(y)) \geq n(x,y)+1$ for all x,y in X and all $m=1,2,\ldots$,

(3) $\lambda(f_m(x),f_m(y)) \leq \alpha\lambda(x,y)$ for all x,y in X and $m=1,2,3,\ldots$.

Next, we define

$$\rho^*(x,y) = \inf\{\sum_{i=1}^{n} \lambda(x_i,x_{i+1}) \mid x_1,\ldots,x_{n+1} \in X \text{ with } x_1 = x \text{ and } x_{n+1} = y,\ n=1,2,\ldots\}$$

for all x,y in X.

It follows from the definition that $\rho^*(x,y) \leq \lambda(x,y) \leq \rho(x,y)$, $\rho^*(x,y) = \rho^*(y,x)$ and $\rho^*(x,x) = 0$ for all $x,y \in X$. We shall show the validity of the triangle inequality for ρ^*. Let $x,y,z \in X$ and $\varepsilon > 0$. From the definition of $\rho^*(x,y)$ and $\rho^*(y,z)$ there exists elements $u_1,u_2,\ldots,u_{n+1}$, $v_1,v_2,\ldots,v_{m+1}$ with $u_1 = x$, $u_{n+1} = y$, $v_1 = y$, $v_{m+1} = z$ and

$$\rho^*(x,y) > \sum_{i=1}^{n} \lambda(u_i,u_{i+1}) - \frac{\varepsilon}{2},\quad \rho^*(y,z) > \sum_{j=1}^{m} \lambda(v_j,v_{j+1}) - \frac{\varepsilon}{2}.$$

Hence, we have $\rho^*(x,y) + \rho^*(y,z) > \sum_{i=1}^{n} \lambda(u_i,u_{i+1}) + \sum_{j=1}^{m} \lambda(v_j,v_{j+1}) - \varepsilon \geq \rho^*(x,z) - \varepsilon$. Since $\varepsilon > 0$ is arbitrary, $\rho^*(x,y) + \rho^*(y,z) \geq \rho^*(x,z)$. In order to prove that ρ^* is a metric on X it remains to show that $\rho^*(x,y) \neq 0$ for $x \neq y$. Let $x,y \in X$ such that $x \neq y$. We may assume that $x \neq x_0$. Let $\varepsilon > 0$, there exists $x_1 = x,x_2,\ldots,x_{n+1} = y \in X$ such that $\rho^*(x,y) > \sum_{i=1}^{n} \lambda(x_i,x_{i+1})-\varepsilon = \sum_{i=1}^{n} \alpha^{n(x_i,x_{i+1})} \rho(x_i,x_{i+1})-\varepsilon$. We shall distinguish two cases.

<u>Case 1</u>. Suppose that $n(x_2) \leq n(x)$ for every finite sequence $x_1 = x,x_2,\ldots,x_{n+1} = y \in X$ such that $\rho^*(x,y) > \sum_{i=1}^{n} \lambda(x_i,x_{i+1})-\varepsilon$. Let

$$N = \min\{n \mid \exists x_1 = x,x_2,\ldots,x_{n+1} = y \in X \text{ such that } \rho^*(x,y) > \sum_{i=1}^{n} \lambda(x_i,x_{i+1})-\varepsilon\}$$

and let $x_1 = x,x_2,\ldots,x_{N+1} = y \in X$ such that $\rho^*(x,y) > \sum_{i=1}^{N} \lambda(x_i,x_{i+1})-\varepsilon$.

We shall show that $n(x_i) \leq n(x)$ for all $i=1,2,\ldots N+1$. If this were not true then $n(x_i) > n(x)$ for some $i \in \{3,4,\ldots,N+1\}$. Let i_0

be the smallest positive integer > 2 such that $n(x_{i_0}) > n(x)$, then $n(x_i, x_{i+1}) \leq n(x)$ for each $i=1,2,\ldots,i_0-1$. Thus

$$\begin{aligned}
\rho^*(x,y) &> \sum_{i=1}^{N} \lambda(x_i,x_{i+1}) - \varepsilon \\
&= \alpha^{n(x,x_2)}\rho(x,x_2) + \alpha^{n(x_2,x_3)}\rho(x_2,x_3) + \ldots \\
&\quad + \alpha^{n(x_{i_0-1},x_{i_0})}\rho(x_{i_0-1},x_{i_0}) + \alpha^{n(x_{i_0},x_{i_0+1})}\rho(x_{i_0},x_{i_0+1}) \\
&\quad + \ldots + \alpha^{n(x_N,x_{N+1})}\rho(x_N,x_{N+1}) - \varepsilon \\
&\geq \alpha^{n(x)}[\rho(x,x_2) + \rho(x_2,x_3) + \ldots + \rho(x_{i_0-1},x_{i_0})] \\
&\quad + \alpha^{n(x_{i_0},x_{i_0+1})}\rho(x_{i_0},x_{i_0+1}) + \ldots \\
&\quad + \alpha^{n(x_N,x_{N+1})}\rho(x_N,x_{N+1}) - \varepsilon \\
&\geq \alpha^{n(x)}\rho(x,x_{i_0}) + \alpha^{n(x_{i_0},x_{i_0+1})}\rho(x_{i_0},x_{i_0+1}) + \ldots \\
&\quad + \alpha^{n(x_N,x_{N+1})}\rho(x_N,x_{N+1}) - \varepsilon \\
&= \alpha^{n(x,x_{i_0})}\rho(x,x_{i_0}) + \alpha^{n(x_{i_0},x_{i_0+1})}\rho(x_{i_0},x_{i_0+1}) + \ldots \\
&\quad + \alpha^{n(x_N,x_{N+1})}\rho(x_N,x_{N+1}) - \varepsilon
\end{aligned}$$

so that $z_1 = x$, $z_2 = x_{i_0}$, $z_3 = x_{i_0+1}, \ldots, z_{N-i_0+2} = x_N$, $z_{(N-i_0+2)+1} = x_{N+1} = y$ is a finite sequence in X with $\rho^*(x,y) > \sum_{i=1}^{N-i_0+2} \lambda(z_i,z_{i+1}) - \varepsilon$ which contradicts the minimality of N as $i_0 > 2$. Therefore $n(x_i) \leq n(x)$ for all $i=1,2,\ldots,N+1$, hence we have

$$\begin{aligned}
\rho^*(x,y) &> \sum_{i=1}^{N} \alpha^{n(x_i,x_{i+1})}\rho(x_i,x_{i+1}) - \varepsilon \\
&\geq \alpha^{n(x)} \sum_{i=1}^{N} \rho(x_i,x_{i+1}) - \varepsilon
\end{aligned}$$

$$\geq \alpha^{n(x)}\rho(x,y) - \varepsilon \;.$$

Since $\varepsilon > 0$ can be arbitrarily small, $\rho^*(x,y) \geq \alpha^{n(x)}\rho(x,y) > 0$.

Case 2. Suppose that there exists a finite sequence $x_1 = x, x_2, \ldots, x_{n+1} = y \in X$ such that (i) $\rho^*(x,y) > \sum_{i=1}^{n} \lambda(x_i, x_{i+1}) - \varepsilon$ and (ii) $n(x_2) > n(x)$. Then

$$\rho^*(x,y) > \sum_{i=1}^{n} \lambda(x_i,x_{i+1}) - \varepsilon$$

$$\geq \lambda(x,x_2) - \varepsilon$$

$$= \alpha^{n(x,x_2)}\rho(x,x_2) - \varepsilon$$

$$= \alpha^{n(x)}\rho(x,x_2) - \varepsilon$$

$$\geq \alpha^{n(x)}\rho(x,f_1^{\ell_1},\ldots,f_k^{\ell_k}[X]) - \varepsilon \quad \text{where}$$

$$\ell_1+\ell_2+\ldots+\ell_k = n(x_2) \quad \text{and} \quad x_2 \in f_1^{\ell_1}\ldots f_k^{\ell_k}[X]$$

$$\geq \alpha^{n(x)}L - \varepsilon$$

where $L = \inf\{\rho(x,f_1^{t_1}\ldots f_q^{t_q}[X]) \mid q \geq 1 \;,\; t_i \geq 0$ and $t_1+t_2+\ldots+t_q > n(x)\}$.

Clearly, L is independent of ε and we shall show that $L > 0$. Since $x \neq x_0$, $\{x_0\} = \limsup_{n\to\infty} f_n[X] = \bigcap_{n=1}^{\infty} \overline{\bigcup_{k\geq n} f_k[X]}$ and $\{\overline{\bigcup_{k\geq n} f_k[X]} \mid n=1,2,\ldots\}$ is a nested family of compact sets, there exists N such that $\overline{\bigcup_{k\geq N} f_k[X]} \subseteq B_{\frac{1}{2}\rho(x,x_0)}(x_0) = \{z\in X \mid \rho(x_0,z) < \tfrac{1}{2}\rho(x,x_0)\}$.

Thus $\rho(x,f_1^{t_1}\ldots f_q^{t_q}[X]) \geq \tfrac{1}{2}\rho(x,x_0)$ for each $q \geq N$ with $t_q \neq 0$. Hence

$$L \geq \min\{\tfrac{1}{2}\rho(x,x_0),\ \inf\{\rho(x,f_1^{t_1}\ldots f_{N-1}^{t_{N-1}}[X]) \mid 0\leq t_i\leq n(x)+1 \text{ for } i=1,2,\ldots,N-1 \text{ and } t_1+t_2+\ldots+t_{N-1}>n(x)\} \;.$$

Since $B = \{(t_1,\ldots,t_{N-1}) \mid 0 \leq t_i \leq n(x)+1$ for $i=1,2,\ldots,N-1$ and $t_1+t_2+\ldots+t_{N-1} > n(x)\}$ is finite and for any $(t_1,t_2,\ldots,t_{N-1})\in B$, $\rho(x,f_1^{t_1}\ldots f_{N-1}^{t_{N-1}}[X]) > 0$, we conclude that $L > 0$. Since

$\rho^*(x,y) > \alpha^{n(x)}L - \varepsilon$ and ε can be arbitrarily small, we have $\rho^*(x,y) \geq \alpha^{n(x)}L > 0$. This completes the proof that ρ^* is a metric on X.

We show that ρ^* generates the topology ζ on X. Since $\rho^*(x,y) \leq \rho(x,y)$ for all x,y in X, we see that the topology ζ on X induced by ρ is finer than that induced by ρ^*. Since (X,ρ) is compact, the identity map $i: (X,\rho) \to (X,\rho^*)$ is a homeomorphism, hence ρ and ρ^* are equivalent metrics.

Finally, we shall show that $\rho^*(f_m(x),f_m(y)) \leq \alpha\rho^*(x,y)$ for all x,y in X and $m=1,2,\ldots$. Let $x,y \in X$. For $\varepsilon > 0$, there exist n and $x_1,x_2,\ldots,x_n \in X$ such that $x_1 = x$, $x_{n+1} = y$ and $\rho^*(x,y) + \varepsilon > \sum_{i=1}^{n} \lambda(x_i,x_{i+1})$. Then for $m \geq 1$.

$$\begin{aligned}\rho^*(f_m(x),f_m(y)) &\leq \sum_{i=1}^{n} \lambda(f_m(x_i),f_m(x_{i+1})) \\ &= \sum_{i=1}^{n} \alpha^{n(f_m(x_i),f_m(x_{i+1}))} \rho(f_m(x_i),f_m(x_{i+1})) \\ &\leq \sum_{i=1}^{n} \alpha^{n(x_i,x_{i+1})+1} \rho(x_i,x_{i+1}) \\ &= \alpha \sum_{i=1}^{n} \alpha^{n(x_i,x_{i+1})} \rho(x_i,x_{i+1}) \\ &< \alpha(\rho^*(x,y) + \varepsilon) .\end{aligned}$$

Since $\varepsilon > 0$ is arbitrary, $\rho^*(f_m(x),f_m(y)) \leq \alpha\rho^*(x,y)$. This completes the proof.

Corollary 2 and Theorem 3 generalize Theorems 1 and 2 of [1] respectively.

Theorem 4. Let (X,ζ) be a countably compact space and $\{f_n : n=1,2,\ldots\}$ be a commuting family of continuous functions on X with a common fixed point x_0 in X such that

(i) $\bigcap_{n=1}^{\infty} \overline{f_k^n[X]} = \{x_0\}$ for $k=1,2,3,\ldots$

and (ii) $\limsup_{n\to\infty} f_n[X] = \{x_0\}$

then for any $\alpha \in (0,1)$ and for each $D \in \mathcal{P}(\zeta)$ there exists $D^* \in P(\zeta)$ such that Card D = Card D* and $\{f_n | n=1,2,\ldots\}$ is an α-contraction with respect to D^*, i.e., $\rho(f_n(x), f_n(y)) \leq \alpha\rho(x,y)$ for all x,y in X, $\rho \in D^*$ and $n = 1,2,3,\ldots$.

Proof: By Theorem 1, for each $D \in \mathcal{P}(\zeta)$ there exists $E \in \mathcal{P}(\zeta)$ such that Card D = Card E and $d(f_n(x), f_n(y)) \leq d(x,y)$ for all x,y in X and for all $d \in E$ and $n = 1,2,3,\ldots$. Let $E = \{d_\lambda | \lambda \in \Gamma\}$. For each $\lambda \in \Gamma$, let R_λ be the equivalence relation on X defined by $xR_\lambda y$ iff $d_\lambda(x,y) = 0$. Denote by $R_\lambda[x]$ the equivalence class containing x and let $X_\lambda = \{R_\lambda[x] | x \in X\}$. Define $\rho_\lambda(R_\lambda[x], R_\lambda[y]) = d_\lambda(x,y)$ for each $R_\lambda[x]$, $R_\lambda[y]$ in X_λ. Then ρ_λ is a metric on X_λ and since (X,ζ) is countably compact, $(X_\lambda, \rho_\lambda)$ is compact. For each $n = 1,2,3,\ldots,$ define $F_{n,\lambda}: X_\lambda \to X_\lambda$ by $F_{n,\lambda}(R_\lambda[x]) = R_\lambda[f_n(x)]$ for each $R_\lambda[x] \in X_\lambda$. Since each f_n is nonexpansive with respect to d_λ, $F_{n,\lambda}$ is well-defined and continuous and in fact $F_{n,\lambda}$ is nonexpansive with respect to ρ_λ. Since $\overline{\bigcap_{k=1}^{\infty} f_n^k[X]} = \{x_0\}$, it is easy to see that (proof in lemma 3 in [3]) $\bigcap_{k=1}^{\infty} F_{n,\lambda}^k[X_\lambda] = \{R_\lambda[x_0]\}$ for each $n=1,2,\ldots$. Since $f_m f_n = f_n f_m$, we see that $F_{n,\lambda}F_{m,\lambda} = F_{m,\lambda}F_{n,\lambda}$. We shall show that $\limsup_{n\to\infty} F_{n,\lambda}[X_\lambda] = \{R_\lambda[x_0]\}$. Let $R_\lambda[x] \in \limsup_{n\to\infty} F_{n,\lambda}[X_\lambda] = \bigcap_{n=1} \overline{\bigcup_{k\geq n} F_{k,\lambda}[X_\lambda]}$. For each $n=1,2,3,\ldots$ there exists $R[x_{k_n}] \in X$ with $k_{n+1} > k_n$ for each n and $\rho_\lambda(R_\lambda[x], F_{k_n}(R_\lambda[x_{k_n}])) < \frac{1}{n}$. Then $d_\lambda(x, f_{k_n}(x_{k_n})) = \rho_\lambda(R_\lambda[x], R_\lambda[f_{k_n}(x_{k_n})]) = \rho_\lambda(R_\lambda[x], F_{k_n,\lambda}(R_\lambda[x_{k_n}])) < \frac{1}{n}$, $\forall\, n=1,2,\ldots$. Since (X,ζ) is countably compact, $(f_{k_n}(x_{k_n}))_{n=1}^{\infty}$ has a cluster point, say $y \in X$. Then y must be in $\limsup_{n\to\infty} f_n[X] = \{x_0\}$, so $y = x_0$, and we have $d_\lambda(x,x_0) = \liminf_{n\to\infty} d_\lambda(x, f_{k_n}(x_{k_n})) = 0$. This shows $R_\lambda[x] = R_\lambda[x_0]$. Hence $\limsup_{n\to\infty} F_{n,\lambda}[X_\lambda] = \{R_\lambda[x_0]\}$. By Theorem 3, for any $\alpha \in (0,1)$, there exists an equivalent metric ρ_λ^* of ρ_λ on X_λ such that

$$\rho_\lambda^*(F_{n,\lambda}(R_\lambda[x]), F_{n,\lambda}(R_\lambda[y])) \leq \alpha\rho_\lambda^*(R_\lambda[x], R_\lambda[y])$$

for all $x,y \in X$ and $n = 1,2,\ldots$.

Define $d_\lambda^*(x,y) = \rho_\lambda^*(R_\lambda[x], R_\lambda[y])$ for x,y in X. Then clearly d_λ^* is an equivalent pseudometric to d_λ on X. Hence

$D^* = \{d^*_\lambda | \lambda \in \Gamma\} \in \mathcal{P}(\zeta)$ and it is clear that $d^*_\lambda(f_n(x), f_n(y)) \leq \alpha d^*_\lambda(x,y)$ for all x,y in X, $n=1,2,\ldots$ and for all $\lambda \in \Gamma$. This completes the proof.

References

[1] Janos, L., A converse of Banach's contraction theorem, Proc. Amer. Math. Soc. 18 (1967), 287-289.

[2] __________, Topological homotheties on compact Hausdorff spaces, Proc. Amer. Math. Soc. 21 (1969), 562-568.

[3] Tan, K.K., Remark on contractive mappings, Bull. Un. Mat. Ital. (4) 9 (1974), 23-27.

Dalhousie University
Halifax, Nova Scotia
Canada

(1) This research was supported by National Research Council of Canada Grants A-3999 and A-8096. It was prepared while the first author was on sabbatical leave from University of Malaya at Dalhousie University.

INVARIANT RENORMING

M.M. Day

0. Introduction

The fixed-point theorems which started this investigation involve some metric condition on the function and some geometric restriction on the domain in which the fixed-point is sought. Here is an example of the kind.

Theorem O.A.: Let K be a bounded closed convex subset of a uniformly convex space B. Then each isometry of K onto K has a fixed point in K.

This and stronger theorems have been proved by brute force but another way to proceed is to apply an old result.

Theorem O.B.: (Brodskiĭ and Mil'man 1948): Let K be a convex closed set in a Banach space B such that K has a unique Čebyšev center $\check{c}$. Then $\check{c}$ is a common fixed-point for all the isometries of K onto K. [See Appendix 3(ii) for definitions.]

These isometries clearly carry the set of Čebyšev centers of K onto itself. To prove Theorem O.A., note that uniform rotundity of N conceals enough compactness to force K to have at least one Čebyšev center and is at the same time a strong enough roundness condition to force that center to be unique.

Not all spaces are uniformly rotund, but there are a wide variety of renorming theorems which allow the introduction of a new norm isomorphic to the original norm but with some additional roundness or smoothness properties. On the other hand, the original isometries of N may not be isometries in the new norm. This raises the problem discussed here.

When can a new isomorphic norm be constructed in N which is invariant under a group G (perhaps the full group) of those transformations of N onto N which are isometries for the old norm?

When this technique succeeds, it will yield new applications of old fixed-point theorems. There are many questions the method will not handle, but the possibility of these applications is my excuse for talking about invariant renorming at a conference about fixed-points.

1. One success story - superreflexive spaces.

R.C. James 1972 has the well-earned credit for identifying and giving many characterizations of superreflexive spaces. (See my book, Day 1973, Chapter VII, Section 4, B, for an outline of this material.)

Define an order relation between normed spaces: N mimics M means that for each finite-dimensional linear subspace E of M there are isomorphisms of E into N which are arbitrarily close to isometries. N is called superreflexive if N mimics only reflexive Banach spaces.

Recall that to say that N is uniformly rotund (the uniformly convex of Clarkson 1936) means that there is a positive function $\delta(\epsilon)$ defined for each $\epsilon > 0$ such that $||x|| \leqq 1 \geqq ||y||$ and $||x-y|| \geqq \epsilon$ together imply that $||x+y|| \leqq 2(1-\delta(\epsilon))$. Clearly a uniformly rotund

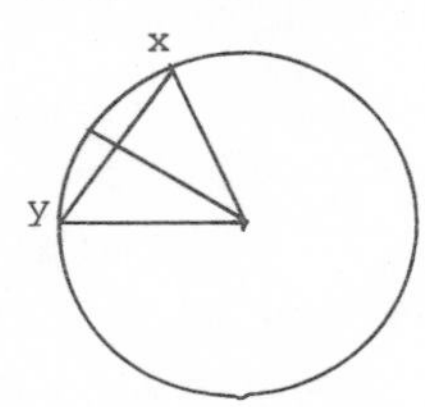

space N mimics only uniformly rotund spaces which have about the same $\delta(\epsilon)$ as N has; so uniformly rotund spaces are superreflexive (by the theorem of Mil'man 1938, Pettis 1939) that each uniformly rotund Banach space is reflexive.

James 1972 gave a different characterization of superreflexive spaces. An n-ϵ-tree in N is a sequence $(x_i,\ 1 \leqq i < 2^{n+1})$ of points of N for which for each $i < 2^n$

(a) $x_i = (x_{2i} + x_{2i+1})/2$ and (b) $||x_{2i}-x_{2i+1}|| \geqq \epsilon$.

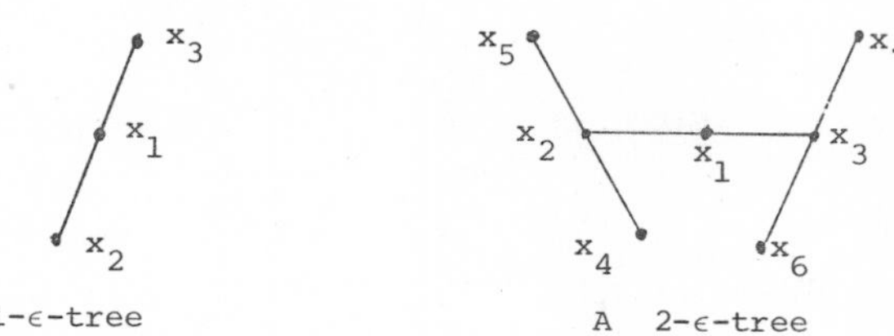

A 1-ϵ-tree   A 2-ϵ-tree

The outreach $r(T)$ of a tree T is $\max\{||x_i|| \mid 1 \leq i < 2^{n+1}\}$. To say that trees grow in N means that for each positive ϵ and each positive integer k there is a positive integer $n(k,\epsilon)$ such that if $n \geq n(k,\epsilon)$ and T is an n-ϵ-tree, then $r(T) \geq k$.

Theorem 1.A. (James 1972): N is superreflexive if and only if trees grow in N .

This was completed soon after

Theorem 1.B. (Enflo 1972): If trees grow in N , then N can be renormed with an isomrophic uniformly rotund norm.

These results show (Enflo 1972) that the three conditions

($\Sigma\rho$) N is superreflexive,

(γ) Trees grow in N ,

<UR> N is isomorphic to a uniformly rotund space,

are equivalent. While generally an isomorphism to improve the rotundity might spoil something else, here all goes well.

Theorem 1.1. Every superreflexive space can be given an isomorphic norm, for example Enflo's norm, which is uniformly rotund and is invariant under all the original isometries of N onto N .

One need only follow Enflo's proof in which he constructs a sequence of functions, the last of which is the new norm, and observe that if T is an isometry of N onto N , then each function constructed is invariant under T .

Theorem O.A. gives a fixed-point theorem with this renorming.

Corollary 1.2. If T is an isometry of a superreflexive Banach space B onto itself which carries a closed bounded convex set K onto K, then T has a fixed point in K .

The Brodskiǐ-Mil'man Theorem O.B. with the result 3.1 that an affine isometry of K onto K can be extended to an isometry of the closed linear span of $K - K$, gives still more from 1.1.

Corollary 1.3. If K is a bounded closed convex set in a superreflexive Banach space B , then there is in K a common fixed point $\check{c}$ of all the affine isometries of K onto K .

Pass to E ; it also is superreflexive so it has a uniformly rotund norm invariant under the original isometries of E onto E . Hence the new unique Čebyšev center $\check{c}$ of K is a common fixed point of all the transformations originally chosen.

2. Special spaces - some successes and some failures.

Brodskii and Mil'man 1948 invented a geometric condition called normal structure whose purpose was to provide uniqueness of Cebysev centers for subsets of K . Garkavi 1962 described a different condition which forces uniqueness of Cebysev centers; (see Appendix 3(ii)).

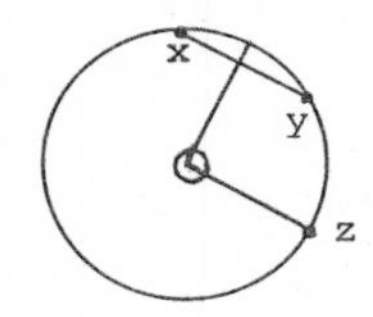

To say that N is uniformly rotund in every direction (symbol: URED) means that for each $\epsilon > 0$ and each z of norm one in N there is a $\delta(\epsilon,z) > 0$ such that if $||x|| \leqq 1 \geqq ||y||$ and $||x-y|| \geqq \epsilon$ and if $x-y = \alpha z$, then $||x+y|| < 2(1-\delta(\epsilon,z))$; that is, the conditions for uniform rotundity are applied to families of parallel chords of the unit ball. This, with the Brodskiǐ-Mil'man Theorem O.B., suggest the value of renorming to get URED if it is possible to have it along with invariance under the old-style rotations. Let us survey the usual list of spaces; for definitions see, for example, Day 1973, Chap. II, Sec. 2.

The spaces $c_0(\Gamma)$. For these spaces all isometries are of the usual form: For all x in $c_0(\Gamma)$ and all γ in Γ ,

$$[Tx](\gamma) = \theta(\gamma)x(\pi(\gamma)) ,$$

where π is a permutation of Γ and $|\theta(\gamma)| = 1$ for all γ in Γ . In $c_0(\Gamma)$ is defined (Day 1955) a rotund norm / / which Rainwater 1969 showed to be locally uniformly rotund. To define this norm let Φ be the set of all one-to-one functions ϕ from ω into Γ ; then define F_ϕ from $c_0(\Gamma)$ into $\ell^2(\omega)$ by: For all n in ω ,

$$[F_\phi x](n) = x(\phi(n))/2^n .$$

Then set

$$/x/ = \sup\{\,||F_\phi x||_2 \mid \phi \text{ in } \Phi\} .$$

(The sup is attained for any ϕ for which the sequence $|x(\phi(n))|$ is non-increasing.) Here is a picture for a cross-section by a two-dimensional coordinate subspace of $c_0(\Gamma)$; the intersecting ellipses have half-axes two and four.

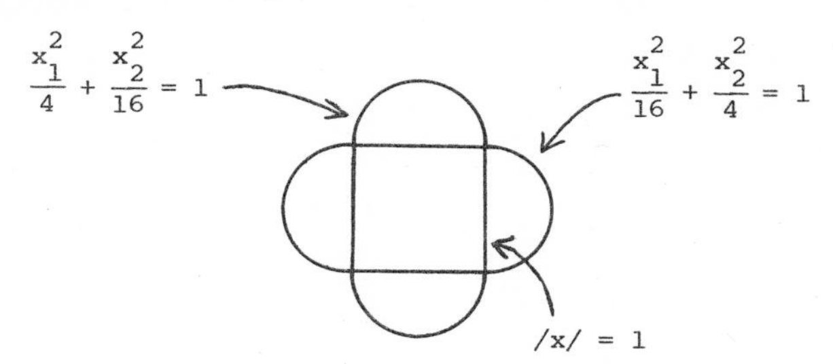

It is obvious that the construction of the norm is invariant under the permutations of Γ or multiplications by functions of absolute value one on Γ , so $c_0(\Gamma)$ has an isomorphic LUR norm which is invariant under all the old-style rotations. Unfortunately, this is the best that is to be expected when Γ is uncountable.

Theorem 2.A. (Day, James, Swaminathan 1971, Zizler 1971): If Γ is uncountable, there is no isomorphic URED norm in $c_0(\Gamma)$.

There are URED norms in $c_0(\omega)$, but the norm of the picture is not one of them. If $c = (\Sigma_i\ 4^{-i})^{-1/2}$, then the sequences $x_n = (c,c,\ldots,c,0,0,\ldots)$ (with n c's) , and $y_n = (-c,c,\ldots,c,0,0,\ldots)$ have $x_n - y_n = (2c,0,0,\ldots) \neq 0$ and $/x_n/ \to 1$, $/y_n/ \to 1$, and $/x_n + y_n/ \to 2$.

Query. Is there an invariant URED norm in $c_0(\omega)$ or in $c(\omega)$?

The spaces $m(\Gamma)$. If Γ is uncountable, the space $m(\Gamma)$ does not have a rotund isomorphic norm (Day 1956), but if Γ is countable, so that $m(\Gamma) = m(\omega)$, then for any single permutation ϕ of ω (say $\phi(n) = n$ for all n in ω) the new norm

$$/x/_\phi = (||x||_\infty^2 + ||F_\phi x||_2^2)^{1/2}$$

is isomorphic to the old norm. That this norm is URED follows from the standard scheme for renorming to get URED.

Theorem 2.B. (Zisler 1971): If M is URED and if F is a one-to-one, continuous linear function from N into M , then an isomorphic URED norm is defined in N by

$$/x/ = (||x||_N^2 + ||Fx||_M^2)^{1/2} .$$

$/\ /_\phi$ is not invariant under many isometries of $m(\omega)$.

When $m(\omega)$ is viewed as the conjugate of $\ell^1(\omega)$, F_ϕ is an adjoint operator. Hence F_ϕ is weak*-to weak* = weak-continuous from $m(\omega)$ into $\ell^2(\omega)$ and hence $/\ /_\phi$ is a conjugate norm in $m(\omega) = \ell^1(\omega)^*$.

Theorem 2.2. There is no conjugate URED, or even rotund, norm in $m(\omega)$ which is invariant under all permutations of ω.

Proof: Let G be a countable group and realize $m(\omega)$ by any one-to-one map between ω and G. If g is an element of G, then the left and right translations in $m(G)$ are defined for g in G and x in $m(G)$ by $\ell_g x(g') = x(gg')$ and $r_g x(g') = x(g'g)$ for all g' in G. Let e be the function which is always equal to 1 on G. Then the set M of means on $m(G)$ is the set of all elements μ in $m(G)^*$ for which $||\mu|| = \mu(e) = 1$ (so $\mu(x) \geqq 0$ if $x \geqq 0$). To say that G is left-amenable is to say that there exists a left-invariant mean; that is, there is a μ in M such that $\ell_g^* \mu = \mu$ for each g in G; that is, the group of all ℓ_g^* have a common fixed point in M. (For lots about amenable semigroups see Day 1957 and 1969 and Greenleaf 1969.)

To say that G is right-stationary means that for each x in $m(G)$ there is at least one constant function in K_x, the w*-closed convex hull of the right orbit $\{r_g x \mid g \text{ in } G\}$. Since G acts transitively on itself, the constant functions are the only common fixed points of all the r_g, g in G.

Theorem 2.C. (Mitchell 1965): G is left-amenable if and only if G is right-stationary.

Now each r_g is an isometry of $m(G)$; each K_x is w*-closed, and is bounded so each K_x has Čebyšev centers relative to the original norm $||\ ||_\infty$ and to any other conjugate norm.

If the new norm were URED, the new Čebyšev center for K_x would be unique and therefore would be a fixed point of all r_g. If the new norm were merely rotund, then the point of K_x of minimum new norm would be unique and therefore would be a point fixed under all

r_g , a constant function in K_x . Mitchell's theorem would then require that all countable groups be amenable. Theorem 1.2 is proved when we observe that this would conflict with a well-known fact.

Theorem 2.D. (von Neumann 1929): There are countable groups, in particular, the free group on two generators, which are not left or right amenable.

Note that if there is no invariant norm in $c_0(\omega)$ which is URED, then Theorem 2.2 could be strengthened; there would be no invariant URED norm in $m(\omega)$ either.

The spaces $L^\infty(\mu)$. If $(S,\mathcal{S},\mu)$ represents a set S , a Borel field $\mathcal{S}$ of subsets of S , and a countably-additive measure μ defined on $\mathcal{S}$, then there is little positive result to give in general because most $L^\infty(\mu)$ spaces have complemented subspaces naturally isometric to $m(\Gamma)$ for some Γ . If $(S,\mathcal{S},\mu)$ is not σ-finite, then an $m(\Gamma)$ with Γ uncountable can be found in $L^\infty(\mu)$; if $L^1(\mu)$ is separable but not finite-dimensional, then $m(\omega)$ can be found in $L^\infty(\mu)$. With suitable attention to completeness of $(S,\mathcal{S},\mu)$, so that $L^\infty(\mu)$ is like $L^1(\mu)^*$, it can be shown from the results from the $m(\Gamma)$ spaces that the same trouble occurs here.

Theorem 2.3. Unless $L^\infty(\mu)$ is finite-dimensional, there is no conjugate norm in $L^\infty(\mu)$ which is URED or even R and also invariant under all measurability-preserving transformations of S into S .

There is a little good news.

Theorem 2.4. If $\mu(S) < \infty$, then the new norm

$$/x/ = (||x||_\infty^2 + ||x||_2^2)^{1/2}$$

is a conjugate norm in $L^\infty(\mu)$ which is URED and is invariant under all measure-preserving transformations of S into S .

The identity transformation which takes x in $L^\infty(\mu)$ to x in $L^2(\mu)$ is the adjoint of the identity tranformation of $L^2(\mu)$ into $L^1(\mu)$, so / / is a conjugate norm. It is URED by Theorem 2.B.

This case covers all compact groups.

Corollary 2.5. If G is a compact group with Haar measure μ , then there is a conjugate norm on $L^\infty(\mu)$ which is URED and is invariant under all left and all right translations in $m(G)$. Hence $L^\infty(G)$ is left- and right-amenable.

Because Haar measure is both left- and right-invariant in G, $||x||_2 = ||r_g x||_2 = ||\ell_g x||_2$ for each x in $L^\infty(\mu)$ and g in G. The other properties of $/\ /$ follow from Theorem 2.3. A variant of Mitchell's theorem 2.C. applies to $L^\infty(\mu)$ since it is the conjugate of $L^1(\mu)$, but of course μ itself defines a left- and right-invariant mean on $L^\infty(\mu)$ and indeed $(\int_G x d\mu)e$ is the only constant function in K_x, the w*-closed convex hull of $\{\ell_h r_g x \mid x$ in $L^\infty(\mu)$ and g, h in $G\}$.

The spaces $C(X)$. Since, by Kakutani's representation theorem for abstract M-spaces (Day 1973, Ch. VI, Sec. 2, Thm. 4) each $L^\infty(\mu)$ space can be regarded as a $C(X)$ space for an appropriate compact Hausdorff X, no general results are to be expected here either. The Banach-Stone Theorem (Day 1973, Ch. V, Sec. 3, Thm. 2) describes the isometries of $C(X)$ in the usual way; $[Tf](x) = \theta(x)f(\pi(x))$ for all x in X and f in $C(X)$, where π is a homeomorphism of X onto X and $|\theta(x)| = 1$ for all x and $\theta \in C(X)$.

The spaces $\ell^1(\Gamma)$ and $L^1(\mu)$. The identity mapping of $\ell^1(\Gamma)$ into $\ell^2(\Gamma)$ is the adjoint of the identity mapping of $\ell^2(\Gamma)$ into $c_0(\Gamma)$, so $\ell^1(\Gamma)$ has a conjugate norm

$$/x/ = (||x||_1^2 + ||x||_2^2)^{1/2}$$

which is URED by Theorem 2.B. and is invariant under all permutations of Γ and all multiplications by all functions whose absolute value is always 1. These are the isometries of $\ell^1(\Gamma)$.

<u>Theorem 2.6</u>. $\ell^1(\Gamma)$ has a URED conjugate norm $/\ /$ which is invariant under all isometries of $||\ ||_1$. This norm is also LUR.

<u>Proof of LUR</u>: Take $/x/ = 1$, $/y_n/ = 1$, and $/x+y_n/ \to 2$. Then

$$||x+y_n||_1^2 + ||x+y_n||_2^2 \to 4 = 2(/x/^2 + /y_n/^2)$$
$$= 2(||x||_1^2 + ||y_n||_1^2 + ||x||_2^2 + ||y_n||_2^2) .$$

Also

(a) $0 \leqq 2(||x||_1^2 + ||y_n||_1^2) - ||x+y_n||_1^2 \to 0$, and

(b) $0 \leqq 2(||x||_2^2 + ||y_n||_2^2) - ||x+y_n||_2^2 \to 0$.

(b) with uniform rotundity of $\ell^2(\Gamma)$ implies $||x-y_n||_2 \to 0$, which in turn implies that for all γ in Γ, $y_n(\gamma) \to x(\gamma)$.

Also (a) implies that $||y_n||_1 \to ||x||_1$. These two facts are enough to make $||x-y_n||_1 \to 0$, so $/ \; /$ is LUR.

When $(S,\mathcal{S},\mu)$ is a non-σ-finite measure space, $L^1(\mu)$ splits into an $\ell^1(\Gamma)$ sum, $P_{\ell^1(\Gamma)} L^1(\mu_\gamma)$, where μ_{γ_1} is atomic and all the other $(S_\gamma,\mathcal{S}_\gamma,\mu_\gamma)$ are homogeneous measure spaces $(I^m,\mathcal{S}_m,\lambda^m)$, m a cardinal number, I the unit interval, λ Lebesgue measure, and $\mathcal{S}_m$ the measurable sets. (See Day 1957a for the use of this work of Kakutani 1941 and Maharam 1942 for the proof that every $L^1(\mu)$ is rotund.) It is not generally true that building substitution spaces preserves URED (see M. Smith 1975, example 4.16), but for this particular norm $/ \; /$ defined above for $\ell^1(\Gamma)$, the space $P_{\ell^1(\Gamma)} B_\gamma$ is URED if all the B_γ are m (Smith 1975, Theorem 4.5). Hence our problem is reduced to testing whether all $L^1(\lambda^m)$ have isomorphic URED norms. At the moment this is not known.

Appendices

(i) Extending isometries.

Theorem 3.i. If K is a convex subset of a normed space N and if T is an affine isometry of K into N, then T determines a linear isometry U of E , the linear hull of $K - K$, into N, so there is an affine isometry V extending T defined by U from the flat hull F of K into N. If $T(K) = K$, then $U(E) = E$ and $V(F) = F$.

Choose x_0 in K, and let $K' = K - x_0$, and define U in K' by $Uy = T(y+x_0) - Tx_0$. Then U is an affine isometry such that $U0 = 0$. The only possible algebraic extension of U works and turns out to be an isometry of E into N. Let $Vx = U(x-x_0) + Tx_0$ to define V in F.

If $TK \subseteq K$, then $UK' \subseteq K - K = K' - K' \subseteq E$, so $UE \subseteq E$. If x is in $K' \subseteq K - K$ and if T takes K onto K, there are u, v in K with $x = Tu-Tv$. Let $s = u-x_0$ and $t = v-x_0$, so $x = Us-Ut$ is in UE. That is $K' \subseteq UE$ so $E \subseteq UE$; that is, $UE = E$ if $TK=K$.

(ii) Čebyšev centers. If E and F are subsets of a normed space N and if x is in N, then $r(x,E) = \inf\{||x-y|| \mid y \text{ in } E\}$, and the Čebyšev radius of E in F, $r(E,F) = \inf\{r(x,E) \mid x \text{ in } F\}$. If there is an x in F such that $r(x,E) = r(E,F)$, then x is called a Čebyšev center of E in F , and the set of all such x is denoted by $\check{C}(E,F)$.

Note that in earlier parts of this paper we were interested in the set $\check{C}(K,K)$.

Theorem 3.ii.A. If E is a bounded subset and F is a weakly [or weakly*] compact subset of the normed space N [the conjugate space N*], then the set $\check{C}(E,F)$ of Čebyšev centers of E in F is w-[w*-] closed and is not empty, and is convex if F is convex

If $(x_n, n \in \omega)$ is a sequence of points of F such that $r(x_n,E) \to r(E,F)$, let (y_m) be a weak [weak*] convergent subnet of (x_n) with $\lim_m y_m = y$. Then $y \in F$ and, for each x in E ,

$$||x-y|| \leqq \lim\inf_m ||x-y_m|| \leqq r(y_m,E) \to r(E,F) = r ,$$

so y is in the set $\check{C}(E,F)$. The other conclusions follow from the fact that $\check{C}(E,F) = \bigcap_{x \in E} U_{x,r}$, where $U_{x,r}$ is the closed ball of radius r about x .

From Theorem 3.ii.A. it follows that every bounded set E in a conjugate space N* has $\check{C}(E,N^*)$ non empty, and every weak [weak*] compact convex set K in N [in N*] has $\check{C}(K,K)$ non-empty, convex, and weak [weak*] compact.

Theorem 3.ii.B. (Garkavi 1962): N is URED if and only if no bounded subset of N has more than one Čebyšev center in N [or in any convex subset F of N] .

If $\check{C}(E,F)$ has in it two points $x \neq y$, then $w = (x+y)/2$ is in $\check{C}(E,F)$ so there exist u_n in E with $\lim_n ||w-u_n|| = r = r(E,F)$. Then $x_n = (u_n-x)/r$ and $y_n = (u_n-y)/r$ have $||x_n+y_n|| \to 2$, $\lim_n ||x_n|| = \lim_n ||y_n|| = 1$, and $x_n-y_n = (x-y)/r \neq 0$, so N is not URED.

If N is not URED, so that there exist x_n , y_n , $z \neq 0$, and $\epsilon > 0$ such that $||x_n|| = ||y_n|| = 1$, $||x_n+y_n|| \to 2$, $x_n-y_n = t_n z$, and $t_n \geqq \epsilon$ for all n , let $u_n = (x_n+y_n)/2$ and let E be the closed convex hull of $\{\pm u_n | n \in \omega\} \cup \{\pm \epsilon z/2\}$. Then $\check{C}(E,E) \supseteq \{tz \mid |t| \leq \epsilon/2\}$.

Lemma 3.ii.C. = Theorem 0.B. (Brodskiǐ-Mil'man 1948): If E is a set which has in it exactly one Čebyšev center $\check{c}$, then $\check{c}$ is a fixed point of every isometry of E onto E .

T an isometry implies that

$$||T\check{c}-Tx|| = ||\check{c}-x|| \leq r(E,E) \quad \text{for all } x \text{ in } E .$$

Since T is onto, the Tx fill up all of E , so $T\check{c}$ is a Čebyšev center for E in E ; by uniqueness, $T\check{c} = \check{c}$.

References

[1] Brodskii, M.S., and Mil'man, D.P., "On the center of a convex set", Dokl. Akad. Nauk SSSR 59 (1948), 837-840.

[2] Clarkson, James A., "Uniformly convex spaces", Trans. Amer. Math. Soc. 40 (1936), 396-414.

[3] Day, Mahlon M., "Strict convexity and smoothness", Trans. Amer. Math. Soc. 78 (1955), 516-528.

[4] __________, "Every L-space is isomorphic to a strictly convex space", Proc. Amer. Math. Soc. 8 (1957a), 415-417.

[5] __________, "Amenable semigroups", Illinois Jour. Math. 1 (1957), 509-544.

[6] __________, "Semigroups and amenability", Semigroups; Proc. of a Symposium at Wayne State Univ., Ed. K. Folley, New York, London Academic Press, (1969), 5-53.

[7] __________, "Normed linear spaces, third edition", Ergebnisse der Math. und ihrer Grenzgebiete, Band 21, Springer-Verlag, Berlin-Heidelberg-New York (1973).

[8] __________, James, R.C. and Swaminathan, S., "Normed linear spaces that are uniformly convex in every direction", Canad. Jour. Math. 23 (1971), 1051-1059.

[9] Enflo, Per, "Banach spaces which can be given an equivalent uniformly convex norm", Israel Jour. Math. 13 (1972), 281-288.

[10] Garkavi, A.L., "The best possible net and the best possible cross-section of a set in a normed space", Izv. Akad. Nauk SSSR Ser. Mat 26 (1962), 87-107; also available in Amer. Math. Soc. Translations, Ser. 2, 39 (1964), 111-132.

[11] Greenleaf, Frederick P., "Invariant means on topological groups and their applications", New York, Van Nostrand Reinhold (1969).

[12] James, R.C., "Some self-dual properties of normed linear spaces", Sympos. on Infinite-dimensional Topology, Ann. of Math. Studies, No. 69 (1972), 159-175.

[13] Kakutani, S., "Concrete representation of abstract (L)-spaces and mean ergodic theorem", Ann. of Math. (2) 42 (1941), 523-537.

[14] Maharam, Dorothy, "On Homogeneous measure algebras", Proc. Nat. Acad. Sci. U.S.A. 28 (1942), 108-111.

[15] Mil'man, D.P., "On some criteria for the regularity of spaces of the type (B), C.R. (Doklady) Acad. Sci. URSS 20 (1938), 243-246.

[16] Mitchell, T., "Constant functions and left-invariant means on semigroups", Trans. Amer. Math. Soc. 119 (1965), 244-261.

[17] von Neumann, J., "Zur allgemeine Theorie des Masses", Fund. Math. 13 (1965), 73-116.

[18] Pettis, B.J., "A proof that every uniformly convex space is reflexive", Duke Math. Jour. 5, 249-253.

[19] Rainwater, J., "Day's norm on $c_0(\Gamma)$", Proc. of Functional Analysis Week (Aarhus 1969), (1969), 46-50. Aarhus: Matem. Inst. Univ.

[20] Smith, Mark, "Directional uniform rotundity in Banach spaces", Thesis, University of Illinois in Urbana-Champaign (1975).

[21] Zisler, V., "On some rotundity and smoothness properties of Banach spaces", Dissertationes Math. (Rozprawy Mat.) No. 87 (1971), Warsaw: Inst. Mat. Polsk. Akad. Nauk.

University of Illinois
Urbana, Illinois

REMARKS AND QUESTIONS CONCERNING NONEXPANSIVE MAPPINGS

Michael Edelstein

1. Brodski and Milman [1] introduced the notion of normal structure for closed convex sets C in a Banach space. By definition, C has normal structure if each bounded convex subset K, having a nonzero diameter, contains a point x such that $\sup\{||x-y||: y \in K\}$ is smaller than the diameter of K. This notion was used by them to show that every weakly compact convex subset of a Banach space, with normal structure, contains a point, the Brodski-Milman center, having the property that all isometries of the set onto itself leave it fixed. In other words, this center is a common fixed point for <u>all</u> isometries of the set onto itself.

For the very construction of the center normal structure is indispensable, but it is an open question whether a weakly compact convex set without normal structure can admit a fixed-point-free isometry onto itself. If such an isometry exists then it cannot be affine, for by a well-known theorem of Markov [10] and Kakutani [7] such a mapping cannot be fixed-point-free. (If the answer to the above question is no then a somewhat unrelated but interesting problem is whether an isometry of a weakly compact convex set onto itself can fail to be affine.) Moreover, the geometry of the orbits under such a mapping must satisfy somewhat restrictive conditions as indicated below.

2. <u>Proposition</u>. Let C be a convex set in a Banach space X and $f : C \to C$ an isometry of C onto itself. Suppose that C has the following property:

(*) If $\phi \neq K \subset C$ is closed convex with $f[K] = K$ then $K = C$.

Then for any $x \in C$,

$$\operatorname{diam}\{f^n(x): n = 0,\pm 1,\ldots\} = \sup\{||x-y||: y \in C\} \quad .$$

Proof. Suppose, for a contradiction, that

$$\delta = \operatorname{diam}\{f^n(x): n=0,\pm 1,\ldots\} < \sup\{||x-y||: y \in C\} \ .$$

Clearly then $\overline{B(x,\delta)} \cap C$ is a proper subset of C containing $\{f^n(x): n=0,\pm 1,\ldots\}$. Since f is an isometry the same is true for $\overline{B(f^m(x),\delta)} \cap C$, $m = 0,\pm 1,\ldots$. Hence $\cap\{\overline{B(f^m(x),\delta)} \cap C: m=0,\pm 1,\ldots\} \neq \phi$. This set however, is a proper closed convex subset of C which is mapped onto itself, against (*). Hence the result.

We note that if $f:C' \to C'$ is an isometry onto a weakly compact convex set C', then for some $C \subset C'$ condition (*) is satisfied (with f replaced by its restriction f/C). Indeed, if $\{K_\alpha\}$ is a family of closed convex subsets of C', linearly ordered by inclusion and such that $f[K_\alpha] = K_\alpha$ then $\cap K_\alpha$ is nonempty and belongs to it. Thus an application of Zorn's Lemma yields a minimal K with $f[K] = K$. Setting $C = K$, the proposition applies.

The following is an obvious consequence of the above proposition.

Corollary. If $f : C \to C$ is an isometry of a closed convex subset of a Banach space onto itself and (*) is satisfied, then for every $\varepsilon > 0$,

$$\operatorname{diam}\{f^n(x): n=0,\pm 1,\ldots\} > \operatorname{diam} C - \varepsilon$$

for some $x \in C$.

Other restrictions.

(i) The unit ball cannot be strictly convex. For if it were then any isometry would have to be affine.

(ii) The isometry cannot be extended to the whole of X for by a well-known result of Mazur and Ulam [11] it would again have to be affine.

(iii) By a more recent result of Mankiewicz [9] the isometry cannot be extended to any open set containing C.

Thus we are led to the first two questions.

Q1. Does there exist a weakly compact convex subset of a Banach space (which is not strictly convex) which admits a fixed point free isometry onto itself?

Q2. Same for an isometry into.

3. Some of the above remarks carry over to nonexpansive mappings. For such mappings Kirk [8] proved that if C is weakly compact convex with normal structure and f maps C into itself then there is always a fixed point. Here affinness cannot be obtained from strict convexity. Thus a counterpart of Q1, a variant of which was already mentioned on a previous occasion by Felix Browder, arises in the following:

Q3. Suppose X is a strictly convex Banach space and $C \subset X$ is weakly compact. Must $f : C \to C$ have a fixed point if it is nonexpansive? If not, what if X is assumed to be locally uniformly convex, reflexive and $C \subset X$ is closed, bounded, convex?

It should be noted that in a uniformly convex Banach space X each convex set has normal structure; so Kirk's theorem applies. Browder [2] and Göhde [5] proved this fact independently.

When considering the possibility of constructing fixed-point-free mappings one is confronted with the task of choosing a suitable Banach space, a suitable subset and a nonexpansive mapping (resp. a suitable isometry). Each orbit then, viewed on its own, is the domain of the restriction of such a mapping. It would seem then that a somewhat simpler task is that of defining mappings on sequences of points first, with the hope of extending them next to the desired domain. In the light of our previous remarks we ask:

Q4. Does every weakly compact convex subset C of a Banach space X have the property that for every $\varepsilon > 0$ it contains a sequence $\{x_n : n=0,\pm 1,\ldots\}$ of diameter greater than $\operatorname{diam} C - \varepsilon$ and such that $x_n \to x_{n+1}$ is an isometry (resp. nonexpansive)?

4. Under certain circumstances one can apply the theory of affine mappings to nonexpansive mappings. This is the case when X is strictly convex and for some $x \in C$ $\{f^n(x): n = 0,1,\ldots\}$ contains a convergent subsequence $f^{n_i}(x) \to y$. Here it is known [3], [4] that $f/\overline{co}\,\{f^n(y)\}$ is an affine isometry and that this last mapping can be extended to an affine isometry of X onto itself. Here the following is open.

Q5. Suppose C is a weakly compact convex subset of a non-strictly-convex Banach space and, for some $x \in C$,

$\{f^n(x): n=0,1,\ldots\}$ contains a convergent subsequence. Must f have a fixed point?

For nonexpansive mappings $f : X \to X$ the fact that $\{f^n(x)\}$ contains a convergent subsequence may fail to imply the existence of a fixed point since orbits may still be unbounded; (counterexamples are known in infinite dimensional Banach spaces [4]). In a finite dimensional strictly convex Banach space this cannot happen. Here the following natural question arises.

Q6. Suppose X is a finite dimensional non-strictly-convex Banach space, $f : X \to X$ is nonexpansive and $\{f^n(x)\}$ has a convergent subsequence for some $x \in X$. Must each orbit be bounded?

References

[1] Brodski, M.S., Milman, D.P., On the center of a convex set (Russian), Dokl. Akad. Nauk SSR 59 (1948), 837-840.

[2] Browder, F.E., Nonexpansive nonlinear operators in a Banach space, Proc. Nat. Acad. Sci. 54 (1965), 1041-1044.

[3] Edelstein, M., On nonexpansive mappings, Proc. Am. Math. Soc. 15 (1964), 689-695.

[4] Edelstein, M., On nonexpansive mappings of Banach spaces, Proc. Cambr. Phil. Soc. 61, (1964), 439-447.

[5] Göhde, D., Zum Prinzip der kontraktiven Abbildung, Math. Nach. 30 (1965), 251-258.

[6] Kakutani, S., Yosida, K., Operator theoretical treatment of Markov's process and mean ergodic theorem, Ann. of Math. 42 (1941), 188-228.

[7] Kakutani, S., Two fixed point theorems concerning bicompact convex sets, Proc. Imp. Acad. Tokyo, 14 (1938), 242-245.

[8] Kirk, W.A., A fixed point theorem for mappings which do not increase distance, Am. Math. Monthly 72 (1965), 1004-1006.

[9] Mankiewicz, P., On extension of isometries in normed linear spaces, Bull. Acad. Polon. Ser. Sci. Math. Astronom. Phys. 20 (1972), 367-371.

[10] Markov, A., Quelques théorèmes sur les ensembles abeliens, Dokl. Akad. Nauk SSR 10 (1936), 311-314.

[11] Mazur, S., Ulam, S., Sur les transformations isométriques d'espaces vectoriels normés, C.R. Acad. Paris, 194 (1932) 946-948.

Dalhousie University
Halifax, Nova Scotia
Canada

FIXED POINT METHODS FOR NODAL PROBLEMS IN DIFFERENTIAL EQUATIONS

G. B. Gustafson*

1. Introduction

The purpose of this paper is to state a model fixed-point theorem associated with the Nodal Problem for the nonlinear boundary value problem (BVP, for brevity)

$$\begin{cases} y''(t) + p(t)y(t)|y(t)|^{\alpha} = 0 \ , \ 0 \leq t \leq 1 \ , \\ y(0) = y(1) = 0 \ , \end{cases} \tag{1.1}$$

that is, the problem of existence of solutions $y_n(t)$ of (1.1) having exactly n zeros on $0 < t < 1$, $n = 0,1,2,\ldots$.

Concerning (1.1), it is assumed that $p(t) > 0$, $p \in C[0,1]$, and $0 < \alpha < \infty$ (Superlinear BVP). The results will in fact apply to more general BVP's than (1.1), and also to certain kinds of singular BVP's. Modifications of the methods and main theorem apply directly to the Sublinear BVP, which corresponds to $-1 < \alpha < 0$.

The formulation of the fixed-point problem can be carried out in a variety of ways. The first kind of formulation appears in the work of Nehari [4], [5]. The presentation of Nehari in abstract terms is to write the nodal problem as a minimization problem for a nonlinear functional $\Phi(y)$ on a certain class A of admissible functions. A second formulation, due essentially to Rabinowitz [6], is as follows: write (1.1) as an integral equation, and seek solutions y of the integral equation in the open set $S_n^+ \subseteq C^1[0,1]$ consisting of functions positive near 0 and having exactly n simple zeros in $(0,1)$. The latter approach is also used by Turner [7], [8].

In contrast, the formulation of the fixed-point problem given in this paper retains the desirable numerical advantages of Nehari's variational formulation, but it allows one to work with other projection methods besides the Ritz method. In particular, the nodal locations explicitly appear in the operator, and they are not disguised in the space of action. This feature, together with the absence of linearization hypotheses in the abstract fixed-point theorem, make it possible to treat sublinear and superlinear nodal problems from the same point of view, both numerically and abstractly.

The transition from the operator-theoretic formulation of the nodal problem to an appropriate fixed-point theorem is governed both by personal tastes and the special features of the nodal problem. The theorem given here is motivated by the superlinear case, and the fact that possible solutions of the nodal problem are one-signed between nodes. Therefore, a cone-theoretic theorem is given which recovers known results for (1.1) in the case $0 < \alpha < \infty$. There is a similar theorem suitable for the sublinear case, $-1 < \alpha < 0$; the formulation of this theorem is left to the reader.

2. Formulation of the Fixed-point Problem.

Let $G(t,s;\, a,b)$ denote the Green's function for the scalar BVP $y''(t) = -f(t)$, $y(a) = y(b) = 0$, then $y(t) = \int_a^b G(t,s;a,b)f(s)\,ds$ (see [2], p. 328).

If $y(t)$ is a solution of (1.1) having exactly n zeros in $(0,1)$, located at $a_1 < a_2 < \dots < a_n$, then the following equations must be satisfied:

$$|y(t)| = \int_{a_i}^{a_{i+1}} G(t,s;a_i,a_{i+1})p(s)\,|y(s)|^{1+\alpha}ds, \quad a_i \le t \le a_{i+1} \, , \tag{2.1}$$

$0 \le i \le n$, where $a_0 \equiv 0$, $a_{n+1} \equiv 1$;

$$0 = \int_{a_i}^{a_{i+1}} \frac{\partial G}{\partial t}\,(a_{i+1},s;\; a_i,a_{i+1})p(s)\,|y(s)|^{1+\alpha}ds \; , \\ + \int_{a_{i+1}}^{a_{i+2}} \frac{\partial G}{\partial t}\,(a_{i+1},s;\; a_{i+1},a_{i+2}(p(s)\,|y(s)|^{1+\alpha}ds \; , \tag{2.2}$$

$0 \le i \le n - 1$.

Conversely, it is routine to verify that a function $y \in C[0,1]$ satisfying (2.1)-(2.2) with $y \not\equiv 0$ is a solution if the BVP (1.1) having exactly n zeros at the locations $a_1,\ldots,a_n$.

The abstract formulation is obtained by setting $x(t) = |y(t)|$ in (2.1)-(2.2) and taking x to belong to the cone (see Krasnosel'skii [3])

$$K \equiv \{x \in C[0,1]: x(t) \geq 0\} .$$

Therefore, one is motivated to introduce the tetrahedron

$$T \equiv \{a \in R^n: a = (a_1,\ldots,a_n),\ 0 \equiv a_0 < a_1 < \ldots < a_n < a_{n+1} \equiv 1\}$$

and operators

$$B(a,x): \bar{T} \times K \to K ,$$
$$F(a,x): \bar{T} \times K \to R^n ,$$

as follows: $B(a,x)$ is the RHS of (2.1) and $F(a,x)$ is the RHS of (2.2), with $|y|$ replaced by x .

Lemma 2.1. The nodal problem (1.1) has a solution iff the pair of operator equations

$$\begin{cases} x = B(a,x) \\ 0 = F(a,x) \end{cases} \tag{2.3}$$

has a solution (a,x) with $a \in T$ and $0 \neq x \in K$. The connection between (2.3) and (1.1) is that $x(t) = |y(t)|$, and $a = (a_1,\ldots,a_n)$ gives the nodal locations.

3. A-priori Bounds.

In this section, several elementary facts are recorded concerning the nodal problem for (1.1). The details are left to the reader, for the most part, however references [1], [2], [3] should provide the techniques needed for the proofs.

Lemma 3.1. Let $0 \leq a < b \leq 1$ and put

$$m[a,b](t) = \min\left\{\frac{t-a}{b-a}, \frac{b-t}{b-a}\right\}, \quad t \in [a,b] . \tag{3.1}$$

If $x(t) \geq 0$ is concave on $[a,b]$ and $x(a) = x(b) = 0$, then

$$x(t) \geq ||x|| \cdot m[a,b](t) , \quad t \in [a,b] , \tag{3.2}$$

where $||x|| = \max\{|x(t)|: t \in [a,b]\}$.

Lemma 3.2. Let $y(t)$ be a solution of any comparison equation of the form

$$y''(t) + [p(t)|y(t)|^\alpha + q(t)]y(t) = 0$$

with $q(t) \geq 0$ and continuous.

If a and b are adjacent zeros of $y(t)$, and $||y|| \equiv \max\{|y(t)|: t \in [a,b]\}$, then:

The function $|y|$ is concave on $[a,b]$ and

$$|y(t)| \geq ||y|| m[a,b](t) . \tag{3.3}$$

The norm $||\cdot||$ satisfies

$$||y|| \leq 4[4\pi^2/(b-a)^2 \min\{p(t): t \in [a,b]\}]^{1/\alpha} . \tag{3.4}$$

For

$$t \in [a,b], |y(t)| \leq \min\{|y'(a)|(t-a), |y'(b)|(b-t)\} . \tag{3.5}$$

If $q(t) = 0$ on $[a, a + \frac{b-a}{3}]$ and $[b - \frac{b-a}{3}, b]$, then

$$\max\{|y'(a)|, |y'(b)|\} \leq \frac{3||y||}{b-a} + \left\{\int_0^1 p(t)dt\right\} ||y||^{1+\alpha} , \tag{3.6}$$

<u>Proof</u>: Relation (3.4) is an eigenvalue estimate, (3.5) is a property of concave functions; (3.6) follows from the mean value theorem.

<u>Lemma 3.3</u>. Let $y(t)$ be a solution of a comparison equation

$$y''(t) + [p(t)|y(t)|^\alpha + q(t)]y(t) = 0 \tag{3.7}$$

with zeros at $0 = a_0 < a_1 < \ldots < a_n < a_{n+1} = 1$. Assume $0 \leq q \in C[0,1]$ and the support of $q(t)$ lies in $\bigcup_{i=0}^{n} J_i$,

$$J_i = \left[a_i + \frac{a_{i+1}-a_i}{3}, a_{i+1} - \frac{a_{i+1}-a_i}{3}\right], \quad 0 \leq i \leq n .$$

If $L \equiv \int_0^1 p(t)dt$, $M_i \equiv \max\{|y(t)|: a_i \leq t \leq a_{i+1}\}$, $H_i \equiv a_{i+1} - a_i$, $N_i \equiv M_i/H_i$, $0 \leq i \leq n$, $D(x) = 3x + Lx^{1+\alpha}$, $p \equiv \min\{p(t): 0 \leq t \leq 1\}$, then for

$$1 \leq i \leq n , \quad N_{i-1} \leq D(N_i) ; \tag{3.8}$$

for

$$0 \leq j \leq n-1 , \quad N_{j+1} \leq D(N_j) ; \tag{3.9}$$

For at least one index k, $H_k \geq \frac{1}{n+1}$, and therefore (3.4) implies

$$\begin{aligned} \max\{|y(t)|: 0 \leq t \leq 1\} &\leq D^n(N_k) \\ &\leq R_n \equiv D^n(4p^{-1/\alpha}(n+1)[2\pi(n+1)]^{2/\alpha}) . \end{aligned} \tag{3.10}$$

Proof: (3.8) and (3.9) follow from (3.5) and (3.6).

Lemma 3.4. Let $y(t)$ be a solution of a comparison equation

$$y''(t) + [p(t)|y(t)|^{\alpha} + q_0(t)]y(t) = 0 \tag{3.11}$$

with $q_0 \in C[0,1]$,

$$|q_0(t)y(t)| \leq r .$$

If $\left(\int_0^1 p(t)dt\right)r^{\alpha} < 7/8$, then the only solution of (3.11) satisfying $|y(t)| \leq r$, $y(0) = y(1) = 0$, is $y(t) \equiv 0$.

4. Geometric Properties of $F(a,x)$.

The mapping F is finite-dimensional and continuous, therefore for any continuous mapping $x: \bar{T} \to C[0,1]$ for which $F(a,x(a)) \neq 0$, $a \in \partial T$, the Brouwer degree $d(F(\cdot,x(\cdot)),T,0)$ is defined.

The purpose of this section is to give some geometric results which insure that $d(F(\cdot,x(\cdot)),T,0) = \pm 1$, for a certain class of functions x.

Definition 4.1. A function $N: \partial T \to R^n \setminus \{0\}$ is called an outer normal to the bounded open set $T \subseteq R^n$ provided for some $a^* \in T$

$$N(a) \cdot (a-a^*) > 0 \quad \text{for} \quad a \in \partial T , \tag{4.1}$$

$$\bar{T} \subseteq \{b \in R^n : N(a) \cdot (b-a) \leq 0\} \quad \text{for each} \quad a \in \partial T . \tag{4.2}$$

Let $c > 0$ be arbitrary and consider the function $k(t;a,c)$ given for $a \in T$ by the relation

$$k(t;a,c) = 2cm[a_i,a_{i+1}](t) , \quad a_i \leq t \leq a_{i+1} , \tag{4.3}$$

$i = 0,\ldots,n$. This function has a natural continuous extension to the closure of T, and we denote it by $k(t;a,c)$ again. It is easily verified that

$$\max\{k(t;a,c): t \in [0,1]\} = c . \tag{4.4}$$

Define a function $N: \partial T \to R^n \setminus \{0\}$ in the following way. Let $\{e_i\}_{i=1}^n$ denote the standard unit vectors in R^n, put $e_{n+1} = 0$,

$$F_0 = \{a \in \partial T: 0 = a_1 = \ldots = a_j < a_{j+1} \text{ for some } j , 1 \leq j \leq n\} ,$$

and define faces $F_1,\ldots,F_n$ of ∂T inductively by the equality

$$F_i = \{a \in \partial T: a_{i-1} < a_i = a_{i+1}\} \setminus \bigcup_{k=0}^{i-1} F_k \quad (1 \leq i \leq n) .$$

Then ∂T is the disjoint union of the faces $F_0,\ldots,F_n$. Define

$$N(a) = e_k - e_{k+1} \quad \text{for} \quad a \in F_k \ , \ 1 \leq k \leq n \ , \tag{4.5}$$

$$N(a) = -e_j \ , \quad \text{for} \quad a \in F_0 \quad \text{satisfying} \quad 0 = a_1 = \ldots = a_j < a_{j+1} \ . \tag{4.6}$$

Lemma 4.2. The function $N\colon \partial T \to R^n\backslash\{0\}$ given by (4.5)-(4.6) is an outer normal to the convex open region T .

Lemma 4.3. Suppose $x\colon \bar{T} \to C[0,1]$ is continuous and $c \in (0,\infty)$. If $a \in \partial T$, $x(a) \in C[0,1]$

$$k(\cdot,a,c) \leq x(a) \ , \tag{4.7}$$

then the outer normal N given by (4.5)-(4.6) and the operator F of section 2 satisfy

$$N(a) \cdot F(a,x(a)) < 0 \ . \tag{4.8}$$

Lemma 4.4. Let $k(a) \geq k(\cdot;a,c)$ for some $c > 0$, then

$$d(-F(\cdot,k(\cdot)),T,0) = 1 \ . \tag{4.9}$$

Proof: Relation (4.9) follows from (4.7), (4.8) and the homotopy $H(a,\lambda) = (1-\lambda)[a-a^*]-\lambda F(a,k(a))$, $a \in \bar{T}$, $a^* \in T$, $0 \leq \lambda \leq 1$.

5. A Fixed Point Theorem.

Let T be a bounded open set in R^n , E a Banach space over the real numbers, K a cone in E , and put

$$M = \{x \in K\colon r \leq ||x|| \leq R\} \ .$$

It will be presupposed that operators $F\colon \bar{T} \times K \to R^n$, $B\colon \bar{T} \times M \to K$, $k\colon \bar{T} \to K$, $h\colon \bar{T} \to K$ are given and we seek to find a solution $(a,x) \in T \times M$ to the pair of operator equations

$$\begin{cases} 0 = F(a,x) \ , \\ x = B(a,x) \ . \end{cases} \tag{5.1}$$

In the case of the nodal problem for (1.1), this solution corresponds to a solution of (1.1) with zeros located at $a_1,\ldots,a_n$. However, the abstract formulation is potentially useful for other kinds of problems.

Theorem 5.1. Under the following assumptions, the operator equations (5.1) have a solution in $\bar{T} \times M$:

The operator B is completely continuous and the operators F, h, k are continuous. (5.2)

If $\tau > 0$ and $x \geq \tau k(a)$ or $x \geq \tau h(a)$, $||x|| \leq R$, $a \in \partial T$, then $F(a,x) \neq 0$. Further, $x = B(a,x)$, $F(a,x) = 0$, $a \in \partial T \rightarrow ||x|| \neq r$. (5.3)

The mappings h and k satisfy for $a \in \bar{T}$: $||k(a)|| < r < R$; $\sup\{||x-B(a,y)||: x \in K, (a,y) \in \bar{T} \times M, ||x|| \leq R, F(a,x) = 0\} < ||h(a)||$. (5.4)

The Brouwer degree $d(-F(\cdot,k(\cdot)),T,0)$ is nonzero. (5.5)

If $(a,y) \in T \times M$, $0 < \lambda \leq 1$, and $y = B(a,y)+\lambda h(a)$, $F(a,y) = 0$, then $||y|| \neq R$. (5.6)

If $(a,x) \in T \times M$, $0 < \lambda \leq 1$, and $x = \lambda B(a,x)+(1-\lambda)k(a)$, $F(a,x) = 0$, then $||x|| \neq r$. (5.7)

Remark 5.2. The operator equations (5.1) on $\bar{T} \times M$ take the form $(a,x) = (a+F(a,x),B(a,x))$, therefore Theorem 5.1 is truly a fixed point theorem.

The ideas behind the formulation and proof of Theorem 5.1 originate from the cone theorems of Krasnoselskii [3] and Leray-Schauder degree computations on the spheres $||x|| \leq r$, $||x|| \leq R$ for specially constructed extensions of F and B. In effect, the conclusion of theorem 5.1 follows by showing that the degree of an extension of the mapping $(a,x) \rightarrow (-F(a,x),x-B(a,x))$ on the annulus $r < ||x|| < R$ has degree equal to the degree in (5.5).

Details of the proof, and other results specifically designed for differential equations, shall appear in a forthcoming paper written jointly by Klaus Schmitt and the present author.

6. Application of the Fixed-point Theorem to the Nodal Problem.

The operator $B(a,x)$ is the composition of a linear integral operator $L(a,u)$ and the Nemytskii operator $x \rightarrow u \equiv p(t)x|x|^{\alpha}$.

The functions $h(a)$, $k(a)$ are constructed from the differential operator $(d/dt)^2$ and the boundary conditions $y(0) = y(1) = 0$, as follows. On the open set $0 = a_0 < a_1 < \cdots < a_{n+1} = 1$, define

$$h(t;a,d) = \begin{cases} \dfrac{2d}{3(d_i-c_i)}\, m[c_i,d_i](t), & c_i \leq t \leq d_i, \\ 0, & \text{otherwise}, \end{cases}$$

where $c_i = a_i + (a_{i+1}-a_i)/3$, $d_i = a_{i+1} - (a_{i+1}-a_i)/3$, $0 \leq i \leq n$. This sawtooth function has norm depending on d and the distances $a_{i+1} - a_i$, $0 \leq i \leq n$. However,

$$h(a) - L(a,h(\cdot;a,d))$$

is continuously extendable to $\bar{T}$ and $d/[12(n+1)] \leq ||h(a)|| \leq d/12$. Define $k(t;a,d_1) = 3(d_i-c_i)h(t)a,d_1)$ on $c_i \leq t \leq d_i$, $k(t;a,d_1) = 0$ otherwise, $0 \leq i \leq n$. Put

$$k(a) = L(a,k(\cdot;a,d_1)) ,$$

then k has a continuous extension to $\bar{T}$.

Select $r > 0$ so small that $(\int_0^1 p(t)dt)r^\alpha < 7/8$ and $r < R_n + 1$, where R_n is given by (3.10). Take $d_1 = r$ in the definition of $k(a)$ and choose d in the definition of $h(a)$ so large that

$$\frac{d}{12(n+1)} > \sup\{||x-B(a,y)||: (a,y) \in \bar{T} \times M , ||x|| \leq R , x \in \bar{K}\} ,$$

where $R \equiv R_n + 1$; here, K is the cone of nonnegative functions and

$$M = \{y \in K: r \leq ||y|| \leq R\} .$$

Both (5.2) and (5.4) are valid by virtue of the above selections. From lemmas 3.1, 4.3, 4.4 it follows that (5.3) and (5.5) hold.

The actual work reduces to the verification of (5.6)-(5.7). To verify (5.6), assume $(a,y) \in T \times M$ and $y = B(a,y) + \lambda h(a)$, $F(a,y) = 0$. Then, $y \geq k(\cdot;a,\tau)$ for some $\tau > 0$ and y is a solution on each $[a_i,a_{i+1}]$ of a comparison equation $y''+[p|y|^\alpha+q]y=0$, where $q(t) = h(t;a,d)/y(t)$. By lemma 3.3, $||y|| \leq R_n < R$, hence (5.6). The proof of (5.7) uses lemma 3.4. The absence of solutions of equations (5.1) on $\partial T \times M$ follows from lemmas 3.1, 4.3.

References

[1] Gustafson, G.B. and Klaus Schmitt, "Nonzero solutions of boundary value problems for second order ordinary and delay-differential equations." J. Diff. Eqtns., 12 (1972), 129-147.

[2] Hartman, P., "Ordinary Differential Equations." Wiley, New York, 1964.

[3] Krasnoselskii, M.A., "Positive solutions of operator equations." Noordhoff, Groningen, The Netherlands, 1964.

[4] Moore, R.A. and Z. Nehari, "Non-oscillation theorems for a class of nonlinear differential equations." Trans. Amer. Math. Soc., 93 (1959), 30-52.

[5] Nehari, Z., "On a class of second order differential equations." Trans. Amer. Math. Soc., 93 (1959), 101-123.

[6] Rabinowitz, P.H., "A nonlinear Sturm-Liouville problem." Japan-United States seminar on ordinary differential equations, Lecture notes in mathematics No. 243, Springer-Verlag New York, New York, 1971.

[7] Turner, R.E.L., "Nonlinear Sturm-Liouville problems." J. Diff. Eqtns., 10 (1971), 141-146.

[8] Turner, R.E.L., "Superlinear Sturm-Liouville problems." J. Diff. Eqtns., 13 (1973), 157-171.

---

*This research was supported by the U.S. Army under Grant No. DAHC-04-74-G-0208.

FIXED POINTS FOR LOCAL RADIAL CONTRACTIONS

R. D. Holmes

1. Introduction

In this talk, we wish to consider the question of the existence of fixed and periodic points for local radial contractions (see §2). The requirement that a mapping be a local radial contraction is, in general, much weaker than that of the more usual "local contraction" (cf. e.g. [5]). However, when the underlying space is compact, it is easy to show (Proposition 1) that such a mapping is, in fact, a uniform local contraction, and, as such, has a finite set of periodic points (M. Edelstein [1]). When the space is not assumed to be compact, a slight modification to an example due to E. Rakotch [6] (§3), exhibits a space, homeomorphic to the real line, and a local radial contraction with no periodic points. We show (§4) that if the space is connected and locally connected, and the mapping is uniformly continuous or is a homeomorphism onto, then the space can be remetrized in such a way that the mapping is a (global) contraction. Hence, if the space is also complete, we obtain a unique fixed point.

I would like to thank the Department of Mathematics at Dalhousie University for their invitation to participate at the Seminar on Fixed Point Theory and its Applications.

2. Definitions and Notation

Let (X,d) be a metric space and let $f\colon X \to X$ be a mapping of X into itself. f is said to be a <u>local radial contraction</u> if there is a fixed real number λ, $0 \leq \lambda < 1$, such that $\forall x \in X$, $\exists \varepsilon_x > 0$, $\ni d(x,y) < \varepsilon_x \Rightarrow d(f(x),f(y)) \leq \lambda d(x,y)$. If this condition is satisfied for some ε, independent of x, then we say f is a <u>uniform local contraction</u>.

If d is a mapping of $X \times X$ into the real line (d need not be a metric on X) then we use $B_d(x,\varepsilon)$ to denote the set $\{y \in X \mid d(x,y) < \varepsilon\}$ and, for $A \subseteq X$, we use $d[A]$ to be the (finite or infinite) supremum of $\{d(x,y) \mid x,y \in A\}$.

3. Preliminary Results.

We first resolve the case in which X is compact. We have

Proposition 1. If f is a local radial contraction on the compact metric space (X,d) then f is a uniform local contraction on (X,d).

Proof: For each $n=1,2,3,\ldots$ define the set $C_n = \{x \in X \mid \exists y \in X \ni d(x,y) \leq \frac{1}{n}$ and $d(f(x),f(y)) \geq \frac{1+\lambda}{2} d(x,y)\}$ where λ is the contraction ratio for f. If, for some n, C_n is empty, then we must have $d(x,y) < \frac{1}{n}$ implying that $d(f(x),f(y)) < \frac{1+\lambda}{2} d(x,y)$ and thus f is a uniform local contraction.

We derive a contradiction from the assumption that C_n is non-empty for all n. If so, then $\{C_n\}$ is a nested family of non-empty closed sets in the compact space X and thus $\cap_{n=1}^{\infty} C_n$ is non-empty. Let $x \in \cap_{n=1}^{\infty} C_n$ and take a fixed $n > \frac{1}{\varepsilon_x}$. Now $x \in C_n$ and thus, by the definition of C_n, there is a $y \in X$ with $d(x,y) \leq \frac{1}{n} < \varepsilon_x$ and $d(f(x),f(y)) \geq \frac{1+\lambda}{2} d(x,y) > \lambda d(x,y)$. This is a contradiction to f being a local radial contraction.

Using Proposition 1 and applying, for example, Theorems 2 and 3 of [1], we have:

Corollary. If f is a local radial contraction on a compact metric space then f has a finite set of periodic points. If, in addition, the space is connected, f has a unique fixed point.

Next we give an example, essentially due to E. Rakotch [6] of a metric space X, homeomorphic to the real line R, and a homeomorphism $f: X \to X$ which is a local radial contraction (in fact, f is a local contraction) which fails to have periodic points.

Example. Let $G = \{(x,y) \in R^2 \mid y = x \sin \frac{\pi}{x}, -1 \leq x < 0\} \cup \{(0,0)\}$ and for any two points (x_1,y_1), (x_2,y_2) on the graph of $y = x \sin \frac{\pi}{x}$, $-1 \leq x < 0$ let $\ell((x_1,y_1), (x_2,y_2))$ be the length of the arc of G between (x_1,y_1) and (x_2,y_2). As the total arc length of this curve is infinite, there exists a monotonically increasing sequence

$\{z_n\}_{n=0,1,2,\ldots}$ of zeros of the function $y = x \sin \frac{\pi}{x}$ such that $z_0 = -1$, $z_1 = -\frac{1}{3}$ and $\ell((z_n,0), (z_{n+1},0)) > 2^{n+1}$ for all $n \geq 0$.

Now set $G_n = \{(x,y) \mid (x,y) \in G,\ x \geq z_n\}$, $Z_n = \{(x,0) \mid -1 \leq x \leq \frac{z_n}{2^n}\}$, $U_n = Z_n \cup \frac{1}{2^n} G_n$, and $X_n = U_n + (n,0)$. Let $\ell_n((x_1,y_1),(x_2,y_2))$ be the arc length between two points in X_n. As the space X we take $(\bigcup_{n=1}^{\infty} X_n) \sim \{(0,0)\}$ in the metric derived from R^2. X is clearly homeomorphic to R. Define the mapping $f: X \to X$ as follows:

$$f(x,y) = \begin{cases} \left(n + \frac{1}{3}(x-n+1)\, \dfrac{1 + \frac{z_{n+1}}{2^{n+1}}}{1 + \frac{z_n}{2^n}},\ 0\right) & (x,y) \in Z_n + (n,0) \\[2ex] \left(n + \frac{1}{3} + \dfrac{z_{n+1}}{3.2^{n+1}} + \dfrac{2^{n+1} + Z_{n+1}}{3\ell((z_n,0),(z_{n+1},0))}\, \ell_n((x,y),(n+\frac{z_n}{2^n},0)),0\right) & (x,y) \in \frac{1}{2^n}(G_n \sim G_{n+1}) + (n,0) \\[2ex] \left(\dfrac{x+n+2}{2},\ \dfrac{y}{2}\right) & (x,y) \in \frac{1}{2^n} G_{n+1} + (n,0) \end{cases}$$

With this definition, f is easily seen to be a homeomorphism and a local $\frac{1}{2}$-contraction, yet each orbit under f extends to infinity.

S.A. Naimpally [5] has constructed a somewhat more complicated example along these lines in which there is a point x and a subsequence $\{f^{n_i}(x)\}$ of the itterates of x converging to a point z, and yet z fails to be a periodic point.

4. Main Results.

In this section we prove the following two results for local radial contractions on connected, locally connected metric spaces.

Theorem 1. Let (X,d) be a connected and locally connected metric space and let g be a homeomorphism of X onto X which is a local radial contraction. Then there exists a metric δ on X, topologically equivalent to d, such that g is a contraction on (X,δ).

Theorem 2. If (X,d) is a connected and locally connected metric space and $g: X \to X$ is a uniformly continuous local radial contraction, then there exists a metric δ on X, topologically equivalent to d, such that g is a contraction on (X,δ) .

The proofs of these theorems is broken into a number of propositions and lemmas. The first uses the construction of Theorem 2.4, page 95 of [2] to derive a metric with certain desirable properties.

Proposition 2. Let (X,d) be a connected and locally connected metric space. There is a metric D on X , topologically equivalent to d , such that:

(4.1) $\forall\, x\in X$ and $r > 0$, $B_D(x,r)$ is connected.

(4.2) (X,D) is complete if (X,d) is complete.

(4.3) If $f\colon X\to X$ is a local radial λ-contraction with respect to d , and n is such that $\lambda^n < \frac{1}{2}$, then f^n is a local radial $2\lambda^n$-contraction on (X,D).

(4.4) If $f\colon X \to X$ is uniformly continuous with respect to d , then it is uniformly continuous with respect to D .

Proof: We define D on X by setting $D(x,y) = \inf\{d[S]\,|\,S$ is a connected set containing x and $y\}$. As X is connected and locally connected, the collection of all bounded, open, and connected sets form a cover, and hence, for any $x,y \in X$, there is a single such set which contains x and y (cf. [3], p. 108). Thus D is well defined. Clearly $D(x,y) = 0 \Longleftrightarrow x,y$, $D(x,y) = D(y,x)$, and

(4.5) $$D(x,y) \geq d(x,y) .$$

To demonstrate the triangle inequality for D , let $x,y,z \in X$. Then, for any $\varepsilon > 0$, there are connected sets S_1 and S_2 with $x,y \in S_1$, $y,z \in S_2$ $D(x,y) + \frac{\varepsilon}{2} > d[S_1]$ and $D(y,z) + \frac{\varepsilon}{2} > d[S_2]$. Therefore $D(x,y) + D(y,z) + \varepsilon > d[S_1] + d[S_2] \geq d[S_1 \cup S_2] \geq D(x,z)$ and, as ε is arbitrary, we conclude that D is a metric.

If $x\in X$, $\varepsilon > 0$ then, as X is locally connected, there is some $\phi(x,\varepsilon) > 0$ such that $d(x,y) < \phi(x,\varepsilon)$ implies that x and y lie in a connected set of d-diameter less than ε . $(d(x,y) < \phi \Longrightarrow D(x,y) < \varepsilon)$. This, together with (4.5), implies the topological equivalence of D and d . (4.2) now follows immediately.

To demonstrate (4.1), let $B \subseteq X$ be connected and $x,y \in B$. Then, using (4.5), we have

$$D(x,y) = \inf\{d[S] \mid x,y \in S \text{ , } S \text{ connected}\} \leq d[B] \leq D[B]$$

and, taking the supremum over $x,y \in B$, we have $D[B] = d[B]$. Let $y \in B_D(x,r)$ for some $x \in X$, $r > 0$. Then $D(x,y) < r$ and there is a connected set S_y such that $x,y \in S_y$ and $D[S_y] = d[S_y] < r$. This implies that $S_y \subseteq B_D(x,r)$ and, taking the union of all S_y's, $y \in B_D(x,r)$ and noting that they all contain x, we have that $B_D(x,r) = \bigcup_{y \in B_D(x,r)} S_y$ is connected.

Now, suppose $f\colon X \to X$ is a local radial λ-contraction and n is such that $\lambda^n < \frac{1}{2}$. Then, for each $x \in X$, there is an ε_x such that $d(x,y) < \varepsilon_x$ implies $d(f^n(x),f^n(y)) \leq \lambda^n d(x,y)$. Suppose now that $D(x,y) < \varepsilon_x$. Then, for any $\mu > 0$ with $D(x,y) + \mu < \varepsilon_x$, there is a connected set S with $x,y \in S$ and $d[S] < D(x,y) + \mu$. Thus $S \subseteq B_d(x,D(x,y)+\mu) \subseteq B_d(x,\varepsilon_x)$ and $f^n[S] \subseteq B_d(f^n(x),\lambda^n D(x,y) + \lambda^n\mu)$. Now, as $f^n[S]$ is connected and contains $f^n(x)$ and $f^n(y)$, we have

$$D(f^n(x),f^n(y)) \leq d[f^n[S]] \leq 2\lambda^n D(x,y) + 2\lambda^n\mu .$$

Letting $\mu \to 0$ we have that $D(x,y) < \varepsilon_x$ implies $D(f^n(x),f^n(y) \leq 2\lambda^n D(x,y)$ and, as $2\lambda^n < 1$, f^n is a local radial contraction on (X,D).

Finally, suppose $f\colon X \to X$ is uniformly continuous, that is, for each $\varepsilon > 0$, there is a $\rho(\varepsilon) > 0$ such that $d(x,y) < \rho$ implies $d(f(x),f(y)) < \varepsilon$. Now, $D(x,y) < \rho$ implies that there is a connected set S containing x and y with $d[S] < \rho$. Thus $d[f[S]] < \varepsilon$ and $D(f(x),f(y)) < \varepsilon$ proving (4.4).

In the next result we show that a certain type of "uniform local connectedness" (4.7) and a certain "contraction condition" (4.8) on the distance between images imply that the mapping is a uniform local contraction.

<u>Proposition 3</u>. Let ε, σ, λ be a real numbers $\varepsilon,\sigma > 0$, $0 \leq \lambda < 1$ and let X be a topological space. Suppose that d is a mapping of $X \times X$ into the real line, continuous in the second variable, such that:

(4.6) $$\forall\, x \in X \text{ , } d(x,x) \leq \lambda\sigma$$

(4.7) $\quad \forall\, x \in X$, $\exists$ a connected set S_x with
$B_d(x,\varepsilon) \subseteq S_x \subseteq \{y \in X \mid d(x,y) \leq \sigma\}$.

Suppose also that f is a continuous mapping of X into X satisfying:

(4.8) $$d(f(x),f(y)) < \sigma \Rightarrow d(f(x),f(y)) \leq \lambda d(x,y) .$$

Then $d(x,y) < \varepsilon$ implies $d(f(x),f(y)) \leq \lambda d(x,y)$.

Proof: Suppose $d(x,y) < \varepsilon$. Then, by (4.7), there is a connected set S_x such that $y \in S_x \subseteq \{z \mid d(x,y) \leq \sigma\}$. Suppose there is a $z \in S_x$ with $\lambda\sigma < d(f(x),f(z)) < \sigma$. Then, by (4.8), $d(f(x),f(z)) \leq \lambda d(x,z) \leq \lambda\sigma < d(f(x),f(z))$ which is a contradiction. Thus, as S_x is connected, f is continuous, and d is continuous in the second variable, we must have either $d(f(x),f(z)) \leq \lambda\sigma$, $\forall\, z \in S_x$ or $d(f(x),f(z)) \geq \sigma$, $\forall\, z \in S_x$. But $x \in S_x$ and $d(f(x),f(x)) \leq \lambda\sigma$ by (4.6) and so the former must be the case. Using the fact that $y \in S_x$ and applying (4.8) we have $d(f(x),f(y)) \leq \lambda d(x,y)$ as required.

Lemma 1. Let (X,d) be a connected metric space and $\Delta\colon X \times X \to R$ be a mapping which satisfies all the axioms for a metric except possibly the triangle axiom. Suppose that $d(x,y) \leq \Delta(x,y)$ for all $x,y \in X$ and that f is a uniform local contraction in terms of Δ . Then there is a metric δ on X (equivalent to d if $d(x,x_n) \to 0$ implies $\Delta(x,x_n) \to 0$) such that f is a contraction on (X,δ) . Moreover, if the metrics are equivalent, the completeness of (X,d) implies the completeness of (X,δ) .

Proof: First, we define a metric μ by setting $\mu(x,y) = \inf\{\Sigma_{i=1}^{k} \Delta(x_i,x_{i-1}) \mid x_i \in X,\ x_0 = x \text{ , and } x_k = y\}$. Clearly $\mu(x,y) \leq \Delta(x,y)$, and as $d(x,y) \leq \Delta(x,y)$ we have $\Sigma_{i=1}^{k} \Delta(x_i,x_{i-1}) \geq \Sigma_{i=1}^{k} d(x_i,x_{i-1}) \geq d(x,y)$ and hence $d(x,y) \leq \mu(x,y)$. That $\mu(x,y) = 0 \Leftrightarrow x = y$ and $\mu(x,y) = \mu(y,x)$ follow from this and the corresponding properties of Δ . Now, if $x,y,z \in X$ and $\varepsilon > 0$, then there exist $\{x_0,x_1,\ldots,x_k\}$ and $\{y_0,y_1,\ldots,y_m\}$ such that $x_0 = x$, $x_k = y = y_0$, $y_m = z$, $\mu(x,y) + \frac{\varepsilon}{2} > \Sigma_{i=1}^{k} \Delta(x_i,x_{i-1})$ and $\mu(y,z) + \frac{\varepsilon}{2} > \Sigma_{i=1}^{m} \Delta(y_i,y_{i-1})$. But then

$$\mu(x,y) + \mu(y,z) + \varepsilon > \Sigma_{i=1}^{k} \Delta(x_i,x_{i-1}) + \Sigma_{i=1}^{m} \Delta(y_i,y_{i-1}) \geq \mu(x,z) .$$

As ε is arbitrary it follows that μ is a metric. (Topological equivalence follows from $d(x,y) \leq \mu(x,y) \leq \Delta(x,y)$.)

Now, suppose f is a local contraction in terms of Δ. That is $\Delta(x,y) < \varepsilon_0$ implies $\Delta(f(x),f(y)) \leq \lambda\Delta(x,y)$ for some fixed $\varepsilon_0 > 0$, $\lambda, 0 \leq \lambda < 1$. If $\mu(x,y) < \varepsilon_0$, let ε be chosen $0 < \varepsilon < \varepsilon_0 - \mu(x,y)$. Then, by the definition of μ, there is a set $\{x = x_0, x_1, \ldots, x_k = y\}$ such that $\lambda\mu(x,y) + \lambda\varepsilon > \Sigma_{i=1}^{k} \lambda\Delta(x_i,x_{i-1}) \geq \Sigma_{i=1}^{k} \Delta(f(x_i),f(x_{i-1})) \geq \mu(f(x),f(y))$. Letting $\varepsilon \to 0$, we see that f is a uniform local contraction on (X,μ).

The required metric δ can now be defined as

$$\delta(x,y) = \inf\{\Sigma_{i=1}^{k}\mu(x_i,x_{i-1}) \mid x_0=x, x_k=y, \text{ and } \mu(x_i,x_{i-1})<\varepsilon_0\} .$$

As in the case of μ, δ is a metric on X (well defined because X is connected). The conditions, $\mu(x,y) \geq \delta(x,y)$ and $\delta(x,y) < \varepsilon_0 \Rightarrow \delta(x,y) = \mu(x,y)$ imply that μ and δ are equivalent metrics.

Again, for each $\varepsilon > 0$, there is a set $\{x_0 = x, x_1, x_2, \ldots, x_k = y\}$ with

$$\lambda\delta(x,y) + \lambda\varepsilon > \Sigma\lambda\mu(x_i,x_{i-1}) \geq \Sigma\mu(f(x_i),f(x_{i-1})) \geq \delta(f(x),f(y)) .$$

and letting $\varepsilon \to 0$, f is seen to be a contraction on (X,δ).

Finally, if (X,d) is complete, then a Cauchy sequence in (X,δ) is Cauchy in (X,μ) and hence in (X,d) and must thus converge in (X,d). The equivalence of the metrics then implies that it converges in (X,δ) and thus (X,δ) is complete.

Lemma 2. If g^n is a contraction on (X,d), then, for each α, $0 < \alpha < 1$, there is a metric δ on X, equivalent to d and complete if d is, such that g is an α-contraction on (X,δ).

Proof: Theorem 1 of [4] states that if $g: X \to X$ satisfies:

(i) $\exists \xi \in X$ such that $g(\xi) = \xi$.

(ii) $g^n(x) \to \xi$ as $n \to \infty$ for all $x \in X$.

(iii) There is an open neighborhood U of ξ such that $g^n[U] \to \{\xi\}$.

Then for each $\alpha \in (0,1)$ there is a metric δ on X such that g is an α-contraction on (X,δ). (X,δ) is complete if (X,d) is.

We proceed to show that (i), (ii) and (iii) are satisfied in this case.

If our contraction g^n does not have a fixed point ξ, then, by the Banach Contraction Theorem, we may adjoin a single point ξ to X which will be the unique fixed point of g^n. In either case $g^i(x) \to \xi$ as $i \to \infty$ for all $x \in X$, and conditions (i) and (ii) are satisfied.

Set $V = B_d(\xi,1)$ and note that $g^{nk}[V] \subseteq B_d(\xi,\lambda^k)$ $k=1,2,3,\ldots$ where λ is the contraction constant of g^n. If we set $U = \cap_{i=0}^{n-1} g^{-i}[V]$ then U is a neighborhood of ξ and, if $0 \leq t < n$, $g^{nk+t}[U] \subseteq g^{nk}[V] \subseteq B_d(\xi,\lambda^k)$ and condition (iii) is satisfied.

An application of the above theorem now yields the desired result.

We are now ready to prove the main theorems.

<u>Proof of Theorem 1</u>: First, apply Proposition 2 to get $f = g^n$ as a local radial λ-contraction on X equipped with a metric D in which every ball is connected.

Now, define $N(x,y)$, $x,y \in X$, to be the smallest non-negative integer for which either

i) $D(f^{-N}(x),f^{-N}(y)) > \lambda D(f^{-N-1}(x),f^{-N-1}(y))$ or

ii) $D(f^{-N}(x),f^{-N}(y)) > 1$.

Note that, as f is an onto mapping, one of these conditions must be satisfied for some N, and thus $N(x,y)$ is well defined. Also $N(f(x),f(y)) \neq 0$ implies that $N(f(x),f(y)) = N(x,y) + 1$.

Define a function $\Delta(x,y)$ by setting

$$(4.9) \qquad \Delta(x,y) = \frac{D(x,y)}{D(f^{-N(x,y)}(x),f^{-N(x,y)}(y))} \max\{1,D(f^{-N(x,y)}(x),f^{-N(x,y)}(y)\}$$

and note that

$$(4.10) \qquad D(x,y) \leq \Delta(x,y) \leq \max\{\lambda^{N(x,y)},D(x,y)\} .$$

As f is both a homeomorphism and a local radial λ-contraction in (X,D), there is, for every $m \geq 1$ and $x \in X$, an $\varepsilon(x,m) > 0$ such that $D(x,y) < \varepsilon(x,m)$ implies

$$D(x,y) \leq \lambda D(f^{-1}(x), f^{-1}(y)) \leq \lambda^2 D(f^{-2}(x),f^{-2}(y)) \leq \ldots$$
$$\ldots \leq \lambda^m D(f^{-m}(x),f^{-m}(y)) .$$

Thus, if we have a sequence $\{x_m\} \subseteq X$ with $D(x_m,x) \to 0$, we must have $\lim_{m\to\infty} N(x,x_m) = \infty$ and, by (4.10), $\Delta(x_m,x) \to 0$. Conversely, $\Delta(x_m,x) \to 0$ implies $D(x_m,x) \to 0$. Clearly $\Delta(x,y) = 0 \iff x = y$ and $\Delta(x,y) = \Delta(y,x)$

so, with the possible exception of the triangle axiom, Δ is a metric on X equivalent to D.

Now, if $\Delta(f(x),f(y)) < 1$, we have, by (4.9), that $N(f(x),f(y)) \geq 1$ and thus $N(f(x),f(y)) = N(x,y) + 1$. Again from (4.9), either

$$\Delta(f(x),f(y)) = D(f(x),f(y)) \leq \lambda D(x,y) \quad \text{by i)}$$

$$\leq \lambda\Delta(x,y)$$

or

$$\Delta(f(x),f(y)) = \frac{D(f(x),f(y))}{D(f^{1-N(f(x),f(y))}(x), f^{1-N(f(x),f(y))}(y))}$$

$$\leq \frac{\lambda D(x,y)}{D(f^{-N(x,y)}(x), f^{-N(x,y)}(y))} = \lambda\Delta(x,y) .$$

In either case, $\Delta(f(x),f(y)) \leq \lambda\Delta(x,y)$.

Next, if $y \in B_\Delta(x,1)$, we have $D(x,y) \leq \Delta(x,y) < 1$ which implies $y \in B_D(x,1)$. Thus $B_\Delta(x,1) \subseteq B_D(x,1)$. Also, if $y \in B_D(x,1)$, $D(x,y) < 1$ and $\lambda^{N(x,y)} \leq 1$, hence, by (4.10), $\Delta(x,y) \leq 1$. Thus $B_D(x,1)$ is a connected set with $x \in B_\Delta(x,1) \subseteq B_D(x,1) \subseteq \{y | \Delta(x,y) \leq 1\}$ and the conditions of Proposition 3 are satisfied (with $d = \Delta$ and $\sigma = \varepsilon = 1$). Hence f is a uniform local contraction in terms of Δ.

If we now apply Lemma 1, f becomes a contraction, and as $f = g^n$, an application of Lemma 2 yields the desired result.

Corollary. If, in addition to the assumptions of Theorem 1, (X,d) is complete, or there is an $x \in X$ with a convergent subsequence of itterates, then g has a unique fixed point $z \in X$, and for all $y \in X$ the sequence $\{g^i(y)\}$ converges to z.

Proof: All constructions preserve completeness.

Proof of Theorem 2: Apply Proposition 2 to get $f = g^n$ as a uniformly continuous local radial α-contraction on (X,D) where D is a metric equivalent to d and every D ball is connected.

Define the function $\Delta(x,y)$ as follows:

$$\Delta(x,y) = \begin{cases} \max\{1, D(f(x),f(y)) + \frac{1-\alpha}{2} D(x,y)\} & \text{if } D(f(x),f(y)) > \alpha D(x,y) \\ D(f(x),f(y)) + \frac{1-\alpha}{2} D(x,y) & \text{if } D(f(x),f(y)) \leq \alpha D(x,y) \end{cases}$$

Then certainly, $\Delta(x,y) \geq \frac{1-\alpha}{2} D(x,y)$ and $\Delta(x,y) \geq D(f(x),f(y))$.

Now, if $D(x_m,x) \to 0$, then from some point on $D(f(x_m),f(x)) \leq \alpha D(x_m,x)$ and $D(f(x_m,f(x)) \to 0$. Hence, $\Delta(x_m,x) = D(f(x_m),f(x)) + \frac{1-\alpha}{2} D(x_m,x) \to 0$. Clearly $\Delta(x,y) = 0 \iff x = y$ and $\Delta(x,y) = \Delta(y,x)$ so, with the possible exception of the triangle axiom, Δ is a metric on X equivalent to D .

Now, $\Delta(f(x),f(y)) < 1$ implies $\Delta(f(x),f(y)) = D(f^2(x),f^2(y)) + \frac{1-\alpha}{2} D(f(x),f(y))$ and that $D(f^2(x),f^2(y)) \leq \alpha D(f(x),f(y))$. Hence

$$\begin{aligned}\Delta(f(x),f(y)) &\leq \alpha D(f(x),f(y)) + \frac{1-\alpha}{2} D(f(x),f(y)) \\ &= \frac{1+\alpha}{2} D(f(x),f(y)) \leq \frac{1+\alpha}{2} \Delta(x,y) .\end{aligned}$$

Next, by uniform continuity, for every $\varepsilon > 0$, there is a $\rho(\varepsilon)$ such that $D(x,y) < \rho(\varepsilon)$ implies $D(f(x),f(y)) < \varepsilon$. Choose $\varepsilon > 0$ and then pick ρ so that $\rho \leq \rho(\varepsilon)$ and $\varepsilon + \frac{1-\alpha}{2} \rho \leq 1$. Suppose $y \in B_\Delta(x, \frac{1-\alpha}{2} \rho)$ and thus $\frac{1-\alpha}{2} D(x,y) \leq \Delta(x,y) < \frac{1-\alpha}{2} \rho$. Hence $y \in B_D(x,\rho)$ and we have $B_\Delta(x, \frac{1-\alpha}{2} \rho) \subseteq B_D(x,\rho)$. Also, if $y \in B_D(x,\rho)$, then $D(x,y) < \rho \leq \rho(\varepsilon)$ and thus $D(f(x),f(y)) < \varepsilon$ which gives us $\Delta(x,y) \leq \max\{1, \varepsilon + \frac{1-\alpha}{2} \rho\} = 1$. Therefore $x \in B_\Delta(x, \frac{1-\alpha}{2} \rho) \subseteq B_D(x,\rho) \subseteq B_\Delta(x,1)$ where $B_D(x,\rho)$ is connected. The conditions of Proposition 3 are satisfied (with $d = \Delta$, $\sigma = 1$, $\varepsilon = \frac{1-\alpha}{2} \rho$, and $\lambda = \frac{1-\alpha}{2}$) . Thus f is a uniform local λ-contraction in terms of Δ .

As in Theorem 1, an application of Lemmas 1 and 2 yields the desired conclusion.

As with Theorem 1, we have:

<u>Corollary</u>. If, in addition to the assumptions of Theorem 2, (X,d) is complete, or there is an $x \in X$ with a convergent sequence of itterates, then g has a unique fixed point $z \in X$, and for all $y \in X$ the sequence $\{g^i(y)\}$ converges to z .

<u>Remark</u>. E. Rakotch [6] proved the related result that if $f\colon X \to X$ is a local contraction on a complete metric space, and if, for some $x \in X$ the points x and $f(x)$ are connected by an arc of finite length, then f has a unique fixed point in X .

References

[1] Edelstein, M., "On Fixed and Periodic Points Under Contractive Mappings", J. Lon. Math. Soc., 37 (1962) 74-79.

[2] Hall, D.W., and G.L. Spencer, "Elementary Topology", New York, 1955.

[3] Hocking, J.G., and G.S. Young, "Topology", Reading, Addison-Wesley, 1961.

[4] Meyers, P.R., "A Converse to Banach's Contraction Theorem ", J. Res. NBS, 71B (1967) 73-76.

[5] Naimpally, S.A., "A Note on Contraction Mappings", Indag. Math., 26 (1964) 275-279.

[6] E. Rakotch, "A Note on α-Locally Contractive Mappings", Bull. Res. Counc. of Israel. Vol. 10, F4 (1962) 188-191.

SOME FIXED POINT RESULTS FOR NONEXPANSIVE MAPPINGS

L. A. Karlovitz

1. Introduction

Let X be a Banach space, let C be a closed bounded convex subset, and let $T: C \to C$ be a nonexpansive mapping, i.e., $||Tx-Ty|| \leq ||x-y||$ for all $x,y \in C$. Our main concern is with the existence of fixed points of T, i.e., $x \in C$ such that $Tx = x$. If X is reflexive and has normal structure (i.e., each closed bounded convex subset D which contains more than one point must contain a point x such that $\sup\{||x-y||: y \in D\} <$ diameter $D = \sup\{||z-y||: z,y \in D\}$) then T has a fixed point in C. [Basic existence theorems for reflexive spaces: Browder [3], Göhde [4], Kirk [8].] This result will also be a corollary of some of the considerations below. If X is not reflexive, T may fail to have a fixed point. The question of whether the positive result holds for all reflexive spaces has remained open in spite of considerable effort. In this note we shall be concerned with some reflexive as well as some nonreflexive spaces. In particular, one of our results fits into the gap between general reflexive spaces and those having normal structure. To a lesser extent we are also concerned with the iterative construction of fixed points.

Our approach is a classical one. Namely, we derive the existence, and in some cases the construction, of fixed points through the study of sequences of approximate fixed points in conjunction with the geometric structure of the spaces. We present two distinct developments: (1) In Section 2 we abstract a geometric property of Hilbert space and show that it is equivalent to the so-called Opial condition for weakly, or weak*, convergent sequences. This yields fixed point results in some new situations including certain nonreflexive spaces such as ℓ_1 and J_0, the interesting new space of R.C. James. (2) In Section 3 we develop a property of general nonexpansive mappings which is expressed in terms of approximate fixed points. This is then used in conjunction with the geometric properties of certain spaces to derive fixed point results. In particular, we derive some fixed point results for certain reflexive spaces which do not have normal structure.

2. A generalized Hilbert space property and its applications.

The basic ideas and results of this section are contained in a recent paper [7] by the author.

In a general Banach space we say (Birkhoff [2], James [5]) that w is orthogonal to v, $w \perp v$, if $||w|| \leq ||w+\lambda v||$ for all scalars λ. In general, $\perp$ is not symmetric. (Indeed, symmetry characterizes Hilbert spaces for dimension strictly greater than 2.) We shall say that the relation $\perp$ is approximately symmetric if for each $x \in X$, $x \neq 0$, and $\varepsilon > 0$ there exists a closed linear subspace $U = U(x,\varepsilon)$ so that

$$U \text{ has finite codimension,} \tag{1}$$

and

$$||u|| \leq ||u+\lambda x|| \text{ for each } u \in U, \ ||u||=1, \text{ and each } \lambda, \ |\lambda| \geq \varepsilon . \tag{2}$$

To see that this is a natural generalization of the Hilbert space case, we note that if $\perp$ is symmetric then we can choose $U = \{u: f(u) = 0\}$ where f is a linear functional such that $||f|| = 1$ and $f(x) = ||x||$. Then $x \perp u$ for each $u \in U$; and, by symmetry, $u \perp x$ for each $u \in U$, which is stronger than (2).

If X is a conjugate space we shall say that the relation $\perp$ is weak* approximately symmetric if in addition to (1) and (2), U can be chosen to be weak* closed. We shall say that the relation is uniformly approximately symmetric (uniformly weak* approximately symmetric) if it is approximately symmetric (weak* approximately symmetric) and (2) is replaced by the stronger condition

$$||u|| \leq ||u+\lambda x||-\delta, \text{ for some } \delta = \delta(x,\varepsilon) > 0, \text{ for each } u \in U$$
$$||u|| = 1, \text{ and each } \lambda, \ |\lambda| \geq \varepsilon . \tag{2'}$$

It can be readily shown that if X has a uniformly convex unit ball, then approximate symmetry and uniform approximate symmetry are equivalent.

Example 1. If X is a Hilbert space then $\perp$ is symmetric and the unit ball is uniformly convex; hence, $\perp$ is uniformly approximately symmetric.

Example 2. In ℓ_1 the relation $\perp$ is uniformly weak* approximately symmetric. To see this, given $x = (\xi_i)$, choose N so large that $\Sigma_1^N|\xi_i| \geq 3||x||/4$. For each $\varepsilon > 0$ let $U(x,\varepsilon) = \{u=(\eta_i): \eta_i=0, i=1, \ldots,N\}$. Then we readily observe that for each $u \in U$, $||u+\lambda x||-||u|| \geq |\lambda|\ ||x||/2$, which implies the assertion.

Example 3. In the important, newly defined, space J_0 (James [6], Lindenstrauss-Stegall [9]) the relation $\perp$ is also uniformly weak* approximately symmetric. The argument is similar, albeit somewhat more involved, to the one for ℓ_1 above.

Further Examples. In both c_0 and L_p , $p \neq 2$, $\perp$ fails to be even approximately symmetric.

These geometric notions are related to fixed points for nonexpansive mappings by the following:

Theorem 1. Let X be a reflexive Banach space (resp., the conjugate of a separable Banach space). Suppose that the relation $\perp$ is uniformly approximately symmetric (resp., uniformly weak* approximately symmetric). Then for each weakly convergent (resp., weak* convergent) sequence $\{x_n\}$ with limit x_∞ we have

$$\liminf||x_n-x_\infty|| < \liminf||x_n-x|| , \tag{3}$$

for each $x \neq x_\infty$.

The converse also holds provided that in the reflexive case we also assume that X is separable.

Remark 1. The relation (3) is now known as Opial's condition. He first noted in [10] its importance for nonexpansive mappings. Namely, in the appropriate setting, it follows from (3) that the limit of a sequence of approximate fixed points must be a fixed point. However, Opial viewed (3) as an outgrowth of Hilbert space structure which is quite different from our present symmetry considerations. His method demands: (i) reflexivity, (ii) a weakly continuous duality mapping, and (iii) a uniformly convex unit ball.

Our main application of Theorem 1 is the following theorem in which the iterative construction as well as the existence of fixed points is considered.

Theorem 2. Let X be a reflexive Banach space (resp., the conjugate of a separable Banach space). Let C be a bounded closed (resp., weak* closed) convex subset of X. Let the mapping $T: C \to C$ be nonexpansive. Suppose that the relation $\perp$ is uniformly approximately symmetric (resp., uniformly weak* approximately symmetric) in X. Then T has a fixed point in C.

Furthermore, if T is asymptotically regular (i.e., if for each $x \in C$, $||T^{n+1}x - T^n x|| \to 0$ as $n \to \infty$) then for each $x \in C$ the sequence $\{T^n x\}$ converges weakly (resp., weak*) to some $x_0 \in C$ and x_0 is a fixed point of T.

The conclusions of this theorem are known for the case that X is a Hilbert space and, more generally, for ℓ_p, $1 < p < \infty$, (Opial [10]). However, other of its applications, including the following, seem to be new.

Corollary 1. Let C be a weak* closed convex bounded subset of ℓ_1 or of J_0. (E.g., the unit ball). Let $T: C \to C$ be a nonexpansive mapping. Then T has a fixed point in C.

3. A property of nonexpansive mappings and its applications.

In this section we develop a property of nonexpansive mappings, which is expressed in terms of approximate fixed points. This is then used in conjunction with the geometric properties of some spaces to derive fixed point results.

Lemma 1. Let X be a Banach space. Let C be a weakly compact convex subset of X. Let $T: C \to C$ be nonexpansive. Suppose that C is minimal in the sense that it contains no proper closed convex subsets which are invariant under T. Let $\{x_n\}$ be a sequence of approximate fixed points, i.e., $x_n \in C$ and $||Tx_n - x_n|| \to 0$. Then for each $x \in C$

$$\lim_n ||x - x_n|| = \underline{\text{diameter } C} . \tag{4}$$

If the space X also has normal structure then (4) immediately implies that C contains only one point. Hence, by an oft-used

Zorn's lemma argument, which establishes the existence of minimal invariant subsets, we derive the following classical result as a direct corollary.

Corollary 2. (Kirk [8]). Let X be a reflexive Banach space. Suppose that X has normal structure. Let C be a closed bounded convex subset, and let $T: C \to C$ be a nonexpansive mapping. Then T has a fixed point in C.

However the main result of this section is the application of Lemma 1 to a space which does not have normal structure. To define this space, let ℓ_2 be renormed according to

$$||x|| = \max\{||x||_2/\sqrt{2}\ ,\ ||x||_\infty\} \tag{5}$$

where $||\cdot||_2$ (resp., $||\cdot||_\infty$) denotes the usual ℓ_2 (resp., ℓ_∞) norm. This is essentially the space which has been discussed in various places in the recent literature, e.g., Belluce, Kirk, Steiner [1]. It is easily seen that the new norm is equivalent to the ℓ_2 norm and that with the new norm ℓ_2 fails to have normal structure. (Indeed, $C = \{x: x = (\xi_i)\ ,\ \xi_i \geq 0\ ,\ ||x||_2 \leq 1\}$ can be used to establish the latter fact.)

Theorem 3. Let $X = \ell_2$ renormed according to (5). Let C be a bounded closed convex subset of X. Let $T: C \to C$ be nonexpansive. Suppose that T is asymptotically regular, i.e., for each $x \in C$, $||T^{n+1}x-T^nx|| \to 0$. Then T has a fixed point in C.

Remark 2. We do not know if the assumption of asymptotic regularity is necessary for the existence of the fixed point. This is still under investigation. Moreover, we feel that, while Theorem 3 serves as an interesting example, this line of investigation should have more general implications. Besides the issue of asymptotic regularity, one would like to apply these methods to more general classes of reflexive Banach spaces. At present, this can be done for spaces whose norms are sufficiently close to (5). However; the more general case is still under investigation.

4. Proofs

Proof of Theorem 1: Suppose that $\perp$ is uniformly (weak*) approximately symmetric. Suppose that x_n converges weakly (weak*) to x_∞ . For given $x \neq x_\infty$ we extract a subsequence $\{x_{n'}\}$ so that $\lim||x_{n'}-x|| = \liminf||x_{n'}-x||$ and so that $\lim||x_{n'}-x_\infty|| = s$ exists. If $s = 0$, (3) follows directly. If $s > 0$, let $w = x_\infty - x$ and $\varepsilon = 1/2s$. By hypothesis, there exists a closed (weak* closed) linear subspace $U = U(w,\varepsilon)$ so that (1) and (2') are satisfied. We may also assume that $w \notin U$. Relation (2') is equivalent to

$$|\lambda| \leq ||w+\lambda u||-|\lambda|\delta \text{ , for some } \delta > 0 \text{ and for each } u \in U \text{ , } ||u|| = 1 \text{ , and each } \lambda \text{ , } |\lambda| \leq 2s \text{ .} \tag{6}$$

Furthermore, by (1), the subspace spanned by w and U has a finite dimensional complement V . Thus for each n ,

$$x_n - x_\infty = \lambda_n w + u_n + v_n \text{ , } u_n \in U \text{ , } v_n \in V \text{ .} \tag{7}$$

Using the finite dimensionality of V , $w \notin U$, and the convergence of $x_{n'} - x_\infty$, we readily observe that $\lambda_{n'}$, $||v_{n'}|| \to 0$, and hence $||u_{n'}|| \to s$. Therefore $||u_{n'}||/1+\lambda_{n'} \leq 2s$ for n' sufficiently large, and by (6) and (7),

$$\begin{aligned} ||x_{n'}-x|| &= ||x_{n'}-x_\infty+x_\infty-x|| = ||(\lambda_{n'}+1)w+u_{n'}+v_{n'}|| \\ &\geq |\lambda_{n'}+1|\,||w+(||u_{n'}||/\lambda_{n'}+1)u_{n'}/||u_{n'}||\,||-||v_{n'}|| \\ &\geq ||u_{n'}||(1+\delta)-||v_{n'}|| \text{ .} \end{aligned}$$

Hence we conclude that $\lim||x_{n'}-x|| \geq \lim||x_{n'}-x_\infty||(1+\delta)$, which implies (3) .

To prove the converse suppose that $\perp$ is not uniformly (weak*) approximately symmetric. Choose $x \neq 0$ and $\varepsilon > 0$ so that there does not exist U satisfying (1) and (2'). Choose a sequence of linear functionals $\{f_n\}$ which are dense in X* (resp., in the space of which X is the conjugate). Let $U_n = \{u: f_j(u) = 0 \text{ , } j=1,\ldots,n\}$. Since (2') is not satisfied, we can choose $u_n \in U_n$, $||u_n|| = 1$, and λ_n , $|\lambda_n| \geq \varepsilon$, so that $||u_n+\lambda_n x||-1 \to \gamma \leq 0$. Clearly $|\lambda_n| \to \infty$ is impossible. Hence, after possible extraction of a subsequence, we can conclude that

$$||u_n+\lambda x|| - 1 \to \gamma \leq 0 \text{ ,} \tag{8}$$

for some λ , $|\lambda| \geq \varepsilon$. From the choice of the f_n it follows that u_n converges weakly (weak*) to 0 . By construction, $||u_n-0|| = 1$; however, by (8), $||u_n-(-\lambda x)|| - 1 \to \gamma \leq 0$. Since $\lambda x \neq 0$ this contradicts (3). This finishes the proof of Theorem 1.

Proof of Theorem 2. By a standard argument, there exists a sequence $\{x_n\}$ in C such that $||Tx_n - x_n|| \to 0$. After possible extraction of a subsequence, we may assume that the sequence converges weakly (weak*) to some $z \in C$. If $Tz \neq z$ then by (3) we have $\liminf ||x_n - z|| < \liminf ||x_n - Tz||$, which is easily seen to contradict the nonexpansiveness of T. Hence $Tz = z$ and the existence of the fixed point is proved.

To prove the second part of the theorem, let $x \in C$ be arbitrarily chosen. Let $x_n = T^n x$, $n = 1,2,\ldots$. Let $\{x_{n'}\}$ be a weakly (weak*) convergent subsequence with limit z. Since, by asymptotic regularity, $||Tx_{n'} - x_{n'}|| \to 0$, the argument of the preceding paragraph yields $Tz = z$. Next we note that $||z - x_{n+1}|| = ||Tz - Tx_n|| \leq ||z - x_n||$. Therefore the entire sequence $\{||x_n - z||\}$ decreases and $\lim ||x_n - z|| = r$ exists; and, by (3), $\liminf ||x_{n'} - w|| > r$ for $w \neq z$. Now suppose that another subsequence $\{x_{n''}\}$ converges weakly (weak*) to $w \neq z$. Repeating the argument we find $\lim ||x_{n'} - w|| = \lim ||x_{n''} - w|| <$ $\liminf ||x_{n''} - z|| = \lim ||x_{n''} - z|| = \lim ||x_{n'} - z||$ which contradicts (3). Therefore the entire sequence $\{x_n\}$ converges to z. This finishes the proof of Theorem 2.

Proof of Lemma 1. Choose $y \in C$ and let $s = \limsup ||y - x_n||$. Let $D = \{x\colon x \in C,\ \limsup ||x - x_n|| \leq s\}$. It is easily seen that D is nonempty closed and convex. It is also invariant under T. This follows from $||Tx - x_n|| \leq ||Tx - Tx_n|| + ||Tx_n - x_n|| \leq ||x - x_n|| + ||Tx_n - x_n||$ and $||Tx_n - x_n|| \to 0$. By the minimality of C, $D = C$. Extract a subsequence $\{x_{n'}\}$ so that $\lim ||y - x_{n'}|| = s'$ exists. Now suppose that there exists $z \in C$ and a subsequence $\{x_{n''}\}$ of $\{x_{n'}\}$ so that $\lim ||z - x_{n''}|| = t \neq s'$. Let $E = \{x\colon x \in C,\ \limsup ||x - x_{n''}|| \leq \min\{t, s'\}\}$. By repeating the argument, we find $E = C$, which contradicts the fact that not both y and z belong to E. Hence for each $x \in C$, $\lim ||x - x_{n'}||$ exists and equals s'.

We now complete the proof by showing that $s' = r = \text{diameter } C$. From this it will follow that $||y - x_{n'}|| \to r$ whenever $\{||y - x_{n'}||\}$ converges, and therefore $||y - x_n|| \to r$ for the entire sequence. Hence, by repeating the argument above with $\{x_{n'}\}$ replaced by the entire sequence $\{x_n\}$, we shall have proved (4).

To this end, consider $F = \{u: u \in C\ ,\ ||u-x|| \leq s' \text{ for each } x \in C\}$. F is nonempty because we can extract a weakly convergent subsequence, again denoted by $\{x_{n'}\}$, with limit z . Since $||x-x_{n'}|| \to s'$ for each $x \in C$ it follows that $||x-z|| \leq s'$ for each $x \in C$; hence $z \in F$. Now if $s' < r$ then $F \subsetneq C$. However, we propose to show that this is a contradiction by showing that F is invariant under T. To see this, we first note that as a consequence of minimality closed convex hull $(TC) = C$. Hence if u is an arbitrary element of C then for given $\varepsilon > 0$ we can choose $v = \Sigma_{i=1}^{k} \lambda_i Tx_i$ so that $x_i \in C$, $\lambda_i > 0$, $\Sigma\lambda_i = 1$, and $||u-v|| \leq \varepsilon$. Now let w be an arbitrary element of F then

$$\begin{aligned} ||Tw-u|| &\leq ||Tw-v|| + ||v-u|| \\ &\leq \Sigma\lambda_i ||Tw-Tx_i|| + ||v-u|| \\ &\leq \Sigma\lambda_i ||w-x_i|| + ||v-u|| \leq s' + \varepsilon \ . \end{aligned}$$

Since $\varepsilon > 0$ and $u \in C$ are arbitrary this proves that $Tw \in F$. Thus F is invariant under T and hence $s' = r$. This finishes the proof of Lemma 1.

Proof of Theorem 3. By a standard Zorn's lemma argument, there exists a closed convex subset D of C which is minimal, in the sense of inclusion, with respect to the property of being invariant under T. If D consists of a single point then this is a fixed point and we are done. The rest of the proof is devoted to showing that D consisting of more than one point leads, via Lemma 1, to a contradiction.

Suppose D consists of more than one point. We may assume that $0 \in D$, and we let diameter $D = r > 0$. Choose $x_0 \in D$ and let $x_n = T^n x_0$, $n = 1,2,\ldots$. We denote the components of x_n by $x_n = (x_{n,\mu})$, $\mu = 1,2,\ldots$. The following 3 propositions are needed.

Proposition 1. For each $x \in D$, $\lim ||x-x_n||_\infty = r$.

Proof. By contradiction. Suppose that for some subsequence, again denoted by $\{x_n\}$, $\lim ||x-x_n||_\infty = r - \delta$ for some $\delta > 0$ and $x \in D$. We may assume that $x \neq 0$. From this and $||0-x_n||_\infty \leq \text{diam } D = r$ it follows that $||x/2-x_n||_\infty \leq r - \delta/4$ for n sufficiently large.

At the same time, by the uniform convexity of $||\cdot||_2$, it follows from $||x-x_n||_2/\sqrt{2}$, $||0-x_n||_2/\sqrt{2} \leq \text{diam } D = r$ that $||x/2-x_n||_2/\sqrt{2} \leq r - \tau$ for some $\tau > 0$.

Hence, for n sufficiently large, $||x/2-x_n|| \leq r - \min\{\tau,\delta/4\}$, which contradicts (4), because by asymptotic regularity $||Tx_n-x_n|| \to 0$.

qed.

<u>Proposition 2</u>. There does not exist a positive integer μ_0 and a positive number δ so that $|x_{n,\mu}| \leq r - \delta$, for all $\mu \neq \mu_0$ and for all n sufficiently large.

<u>Proof</u>. By contradiction. Suppose μ_0 and δ exist. Choose ε, $0 < \varepsilon < r/4, \delta$. It follows from Prop. 1 that we can select a subsequence, again denoted by $\{x_n\}$, so that $||x_n-0||_\infty \geq r - \varepsilon$ for all n and so that $||x_n-x_m||_\infty \geq r - \varepsilon$ whenever $n \neq m$. Since all elements belong to ℓ_2, for each pair n,m, $n \neq m$, we can choose an integer $\mu(n,m)$ so that

$$r - \varepsilon \leq |x_{n,\mu(n,m)} - x_{m,\mu(n,m)}| . \tag{9}$$

Since $|x_{k,\mu}| \leq r - \delta$, $\mu \neq \mu_0$, and $\varepsilon < \delta$ it follows from $||x_k||_\infty \geq r - \varepsilon$, all k, that $|x_{k,\mu_0}| \geq r - \varepsilon$. Hence, by diam $D = r$, either $|x_{n,\mu_0}-x_{m,\mu_0}| \leq \varepsilon$ or $|x_{n,\mu_0}-x_{m,\mu_0}| \geq 2r - 2\varepsilon$. The second case contradicts diam $D = r$ by the choice of ε. Hence the first case must hold. Whence, by (9), we conclude that $\mu(n,m) \neq \mu_0$ because $r - \varepsilon \not\leq \varepsilon$. Since $\mu(n,m) \neq \mu_0$ we have by assumption that $|x_{n,\mu(n,m)}|$, $|x_{m,\mu(n,m)}| \leq r - \delta$. From this and (9) we readily calculate that

$$|x_{n,\mu(n,m)}| , |x_{m,\mu(n,m)}| \geq \delta - \varepsilon . \tag{10}$$

By a similar calculation it follows from diam $D = r$ and $\varepsilon \leq r/4$ that

$$\mu(n,m) = \mu(n,k) \Rightarrow \mu(m,k) \neq \mu(n,m) . \tag{11}$$

For each n let $Z_n = \{\mu: \mu = \mu(n,m)$ for some m, $m \neq n\}$. Since $||x_n||_2 \leq \sqrt{2r}$ it follows from (10) that there exists an integer p so that

$$\text{cardinality } Z_n \leq p < \infty \quad \text{for} \quad n = 1,2,\ldots . \tag{12}$$

Let $n_1 = 1$. For each $\mu \in Z_1$ consider the set $\{m: \mu(n_1,m) = \mu\}$. Since Z_1 is finite it follows that for some $\mu_1 \in Z_1$ the set $\Gamma_1 = \{m: \mu(1,m) = \mu_1\}$ is infinite. Choose $n_2 \in \Gamma_1$. For $m \in \Gamma_1$, $m \neq n_2$, since $\mu(1,n_2) = \mu(1,m) = \mu_1$, it follows from (11) that $\mu(n_2,m) \neq \mu_1$. Hence we can repeat the argument and choose $\mu_2 \in Z_{n_2}$

so that $\mu_2 \neq \mu_1$ and so that $\Gamma_2 = \{m: \mu(n_2,m) = \mu_2 , m \in \Gamma_1\}$ is infinite. We proceed by induction. Suppose that $\mu_1,\ldots,\mu_k$ and $\Gamma_1 \supset \ldots \supset \Gamma_k$, Γ_k infinite, have been chosen. Choose $n_{k+1} \in \Gamma_k$. For $m \in \Gamma_k$, $m \neq n_{k+1}$, since $\mu(n_i,n_{k+1}) = \mu(n_i,m) = \mu_i$ for $1 \leq i \leq k$, it follows from (11) that $\mu(n_k,m) \neq \mu_i$, $1 \leq i \leq k$. Hence, by the same argument, we can choose $\mu_{k+1} \neq \mu_1,\ldots,\mu_k$ so that $\Gamma_{k+1} = \{m: \mu(n_{k+1},m) = \mu_{k+1} , m \in \Gamma_k\}$ is infinite. Now consider x_k such that $k \in \Gamma_{2p}$. Then by construction $\mu(k,n_i) = \mu_i$, $i = 1,\ldots,2p$. Hence, by (10), $|x_{k,\mu_i}| \geq \delta-\varepsilon$ for $i=1,\ldots,2p$ which contradicts (12) because the μ_i are all distinct. q.e.d.

We shall say that an element x_n of the sequence $\{x_n\}$ is an ε-<u>double</u>, with <u>support</u> $\{\mu_1,\mu_2\}$, if $\mu_1 \neq \mu_2$ and $|x_{n,\mu_1}|,|x_{n,\mu_2}| \geq r - \varepsilon$ for some $\varepsilon > 0$.

<u>Proposition 3</u>. For each $\varepsilon > 0$ and each integer N there exists n so that $n \geq N$ and x_n is an ε-double.

<u>Proof</u>: Given $\varepsilon > 0$ and N we choose M , $M \geq N$, so that $||x_n||_\infty \geq r - \varepsilon/2$ and so that $||x_n-x_{n+1}||_\infty \leq \varepsilon/2$ for all $n \geq M$. The former is possible by Prop. 1 and the latter by the asymptotic regularity of T . Choose μ_0 so that $|x_{M,\mu_0}| \geq r - \varepsilon/2$.

<u>Case 1</u>: $|x_{n,\mu_0}| \geq r - \varepsilon/2$ for all $n \geq M$. <u>Case 2</u>: There exists $m > M$ so that $|x_{m,\mu_0}| < r - \varepsilon/2$. We first consider Case 1. By Prop. 2 we cannot have $|x_{n,\mu}| \leq r - \varepsilon$ for all $\mu \neq \mu_0$ and all $n \geq M$. Hence for some $n \geq M$ we can find μ_1 so that $\mu_1 \neq \mu_0$ and so that $|x_{n,\mu_1}| > r - \varepsilon$. Hence x_n is the desired ε-double with support $\{\mu_0,\mu_1\}$. Now we consider Case 2. Choose k , $M \leq k < m$ so that k is the largest integer less than m for which $|x_{k,\mu_0}| \geq r-\varepsilon/2$. Then, by $||x_n||_\infty \geq r - \varepsilon/2$, there exists $\mu_1 \neq \mu_0$ so that $|x_{k+1,\mu_1}| \geq r - \varepsilon/2$. However, by $||x_k-x_{k+1}||_\infty \leq \varepsilon/2$, $|x_{k+1,\mu_0}| \geq r - \varepsilon$, and so x_{k+1} is the desired ε-double with support $\{\mu_0,\mu_1\}$. q.e.d.

We now conclude the proof. Choose ε , $0 < \varepsilon < r/64$. Choose N so large that $||x_n||_\infty \geq r - \varepsilon/2$ (by Prop. 1) and $||x_n-x_{n+1}||_\infty \leq \varepsilon/2$ (by asymptotic regularity) for all $n \geq N$. Choose n , $n \geq N$, so that x_n is an ε-double with support $\{\mu_1,\mu_2\}$ (by Prop. 3). Now consider any other ε-double x_m with support $\{\mu_3,\mu_4\}$. We note that $\{\mu_1,\mu_2\} \cap \{\mu_3,\mu_4\} = \phi$ is impossible. For otherwise a straightforward calculation shows that $||x_n-x_m||_2 > \sqrt{2}\, r$, which contradicts diam $D = r$. A similar calculation shows that if $\{\mu_1,\mu_2\} = \{\mu_3,\mu_4\}$ then $||x_n-x_m||_\infty \leq r/4$. Thus if x_m is an ε-double with support $\{\mu_3,\mu_4\}$ and with $||x_n-x_m||_\infty \geq r - \varepsilon/2$ then $\{\mu_1,\mu_2\} \cap \{\mu_3,\mu_4\} = \phi$ and $\{\mu_1,\mu_2\} \neq \{\mu_3,\mu_4\}$. Such x_m exist by virtue of Propositions 1 and 2. Hence the integer m^* is well defined by: $n \leq m^*$, x_{m^*} is an ε-double with support $\{\mu_3,\mu_4\}$ which satisfies $\{\mu_1,\mu_2\} \cap \{\mu_3,\mu_4\} \neq \phi$ and $\{\mu_1,\mu_2\} \neq \{\mu_3,\mu_4\}$, and m^* is the smallest integer with these properties. Now we replace n by n^* which is defined by: $n^* \leq m^*$, x_{n^*} is an ε-double whose support $\{\mu_1,\mu_2\}$ is the same as that of x_n, and n^* is the largest integer with these properties. We may assume the $\mu_1 = \mu_3$ and $\mu_2 \neq \mu_4$. A straightforward calculation shows that, by virtue of $||x_{n^*}|| \leq \sqrt{2r}$, $|x_{n^*,\mu_4}| \leq r/4$. Hence by $||x_i-x_{i+1}||_\infty \geq r - \varepsilon/2$, for $i \geq N$, it follows that $n^* < m^* - 1$. We next note that for each i, $n^* < i < m^*$, x_i is <u>not</u> an ε-double. For if it were with support, say, $\{\mu_5,\mu_6\}$ then by the argument above $\{\mu_1,\mu_2\} \cap \{\mu_5,\mu_6\} \neq \phi$; and by the construction of n^*, $\{\mu_1,\mu_2\} \neq \{\mu_5,\mu_6\}$, which contradicts the minimality of m^*. By $||x_i||_\infty \geq r - \varepsilon/2$, for each i, $n^* \leq i < m^*$, there exists $\mu(i)$ so that $|x_{i,\mu(i)}| \geq r - \varepsilon/2$. If for one such i $\mu(i) \neq \mu(i+1)$ then, by virtue of $||x_i-x_{i+1}||_\infty \leq \varepsilon/2$, it follows that x_i and x_{i+1} are both ε-doubles, which is a contradiction. Thus $\mu(i) = \mu_0$, for some μ_0 and all i, $n^* \leq i < m^*$. Now suppose that $\mu_0 \neq \mu_1$. Then, since the support of x_{n^*} is $\{\mu_1,\mu_2\}$, it follows from $||x_{n^*}||_2 \leq \sqrt{2r}$ that $\mu_0 = \mu_2$. From this and $||x_i-x_{i+1}||_\infty \leq \varepsilon/2$ it follows that x_{n^*+1} is an ε-double with support $\{\mu_1,\mu_2\}$, which is a contradiction. Thus $\mu_0 = \mu_1$ and

$$|x_{i,\mu_1}| \geq r - \varepsilon/2 \;,\; i = n^*,\ldots,m^*-1 \;. \tag{13}$$

Now from $|x_{n^*,\mu_4}| \leq r/4$, $|x_{m^*,\mu_4}| \geq r - \varepsilon$ and $||x_i-x_{i+1}||_\infty \leq \varepsilon/2$, $i \geq N$, it follows that we can choose i^*, $n^* < i^* < m^*$ so that

$$r - 3\varepsilon \leq (\mathrm{sgn}(x_{m^*,\mu_4}))x_{i^*,\mu_4} \leq r - 5\varepsilon/2 \;. \tag{14}$$

Thus using (13) and $||x_{i*}||_2 \leq \sqrt{2}r$ we readily calculate

$$\sum_{\mu \neq \mu_1,\mu_4} |x_{i*,\mu}|^2 < 7r\varepsilon \tag{15}$$

We can choose M so that $M > N$ and $||x_j - x_k||_\infty \geq r - \varepsilon/2$ for $n^* \leq j \leq m^*$ and $k \geq M$ (Prop. 1). Now by repeating the construction above we can select ε-doubles x_{n**}, $M \leq n^{**} < m^{**}$, with supports $\{\eta_1,\eta_2\}$ and $\{\eta_3,\eta_4\}$ so that $\eta_1 = \eta_3$ and $\eta_2 \neq \eta_4$. Moreover, by the construction we can choose i^{**}, $n^{**} < i^{**} < m^{**}$ so that x_{i**} satisfies: $|x_{i**,\eta_1}| \geq r - \varepsilon/2$ and

$$r - 3\varepsilon \leq (\operatorname{sgn}(x_{m**,\eta_4}))x_{i**,\eta_4} \leq r - 5\varepsilon/2 . \tag{16}$$

Thus, as in (14), we calculate

$$\sum_{\eta \neq \eta_1,\eta_4} |x_{i**,\eta}|^2 < 7r\varepsilon . \tag{17}$$

Next we note that

$$\{\mu_1,\mu_i\} \cap \{\eta_1,\eta_j\} \neq \phi \quad \text{and} \quad \{\mu_1,\mu_i\} \neq \{\eta_1,\eta_j\} \quad i,j=2,4 \tag{18}$$

To see this we repeat the arguments above. For example, if $\{\mu_1,\mu_2\} \cap \{\eta_1,\eta_3\} = \phi$ then we readily calculate $||x_{n*} - x_{n**}||_2 > \sqrt{2}r$, which contradicts $\operatorname{diam} D = r$. Similarly, if $\{\mu_1,\mu_2\} = \{\eta_1,\eta_2\}$ then $||x_{n*} - x_{n**}||_\infty \leq r/4$, which contradicts $||x_{n*} - x_{n**}||_\infty \geq r - \varepsilon/2$. Since $\mu_1 = \mu_3 \neq \mu_2$, μ_4; $\mu_2 \neq \mu_4$, and $\eta_1 = \eta_3 \neq \eta_2$, η_4; $\eta_2 \neq \eta_4$ it follows readily from (18) that the only possible arrangement implies

$$\mu_1 = \eta_1 \quad \text{and} \quad \mu_4 \neq \eta_4 .$$

Finally we note that by $\operatorname{diam} D = r$ and $|x_{m*,\mu_4}|, |x_{m**,\eta_4}| \geq r - \varepsilon$ we have $(\operatorname{sgn}(x_{m*,\mu_4}))x_{i**,\mu_4} \geq -\varepsilon$ and $(\operatorname{sgn}(x_{m**,\eta_4}))x_{i*,\eta_4} \geq -\varepsilon$. Combining this with (14) and (16) we find

$$|x_{i*,\mu_4} - x_{i**,\mu_4}| , |x_{i*,\eta_4} - x_{i**,\eta_4}| \leq r - 3\varepsilon/2 . \tag{19}$$

By $\operatorname{diam} D = r$, $\mu_1 = \eta_1$, and $|x_{i*,\mu_1}|$, $|x_{i**,\eta_1}| \geq r - \varepsilon/2$, we have

$$|x_{i*,\mu_1} - x_{i**,\mu_1}| \leq \varepsilon/2 . \tag{20}$$

By virtue of (15) and (17) and $\varepsilon < r/64$ we have for $\mu \neq \mu_1,\mu_4,\eta_4$

$$|x_{i*,\mu}-x_{i**,\mu}| \leq 2\sqrt{7r\varepsilon} < r - \varepsilon . \qquad (21)$$

Thus, by (19), (20) and (21), $||x_{i*}-x_{i**}||_\infty < r - \varepsilon$ which contradicts $||x_{i*}-x_{i**}||_\infty \geq r - \varepsilon/2$. This finishes the proof of Theorem 3.

References

[1] Belluce, L.P., W.A. Kirk and E.F. Steiner, Normal structure in Banach spaces, Pacific J. Math. 26 (1968) 433-440.

[2] Birkhoff, G., Orthogonality in linear metric spaces, Duke Math. J. 1 (1935) 169-172.

[3] Browder, F.E., Nonexpansive nonlinear operators in a Banach space, Proc. Nat. Acad. Sci. 54 (1965), 1041-1044.

[4] Göhde, D., Über Fixpunkte bei stetigen Selbstabbildungen mit kompakten Iterierten, Math. Nach. 28 (1964) 45-55.

[5] James, R.C., Orthogonality and linear functionals in normed linear spaces, Trans. Amer. Math. Soc. 61 (1947) 265-292.

[6] James, R.C., A separable somewhat reflexive Banach space with non-separable dual, Bull. Amer. Math. Soc. 80 (1974) 738-743.

[7] Karlovitz, L.A., On nonexpansive mappings, Tech. Note BN-805, University of Maryland, (1974), to appear Proc. Amer. Math. Soc.

[8] Kirk, W.A., A fixed point theorem for mappings which do not increase distance, Amer. Math. Monthly, 72 (1965) 1004-1006.

[9] Lindenstrauss, J. and C. Stegall, Examples of spaces which do not contain ℓ_1 and whose duals are not separable, to appear.

[10] Opial, Z., Weak convergence of the sequence of successive approximations for nonexpansive mappings, Bull. Amer. Math. Soc. 73 (1967) 591-597.

SOME FIXED POINT THEOREMS FOR SEMIGROUPS OF SELFMAPPINGS

Mo Tak Kiang

In the following some fixed point theorems for linearly ordered semigroups of selfmappings are discussed.

0. Introduction

The concept of a linearly ordered semigroup and that of an Archimedean linearly ordered semigroup were first introduced in [2]. They were motivated by the observation of certain properties possessed by a flow (see [1]).

Definition. Let (X,d) be a metric space and $\mathcal{F}$ a semigroup of selfmappings on X . $\mathcal{F}$ is called linearly ordered if it satisfies: $\forall f, g \in \mathcal{F}$, either $\mathcal{F}f \subseteq \mathcal{F}g$ or $\mathcal{F}g \subseteq \mathcal{F}f$. In the case when $\mathcal{F}f \subseteq \mathcal{F}g$, f is said to follow g and this is denoted by $f \geq g$.

Definition. A linearly ordered semigroup $\mathcal{F}$ is said to be Archimedean at g , (for $g \in \mathcal{F}$ and $g \neq Id$) , if $\forall f \in \mathcal{F}$ with $g \leq f$, $\exists n \in \mathbb{N}$ such that $g^n \geq f$.

Definition. A linearly ordered semigroup $\mathcal{F}$ is said to be Archimedean if $\mathcal{F}$ is Archimedean at each $g \in \mathcal{F}$ where $g \neq Id$.

It is clear (see [2]) that a flow and also any semigroup of mappings with a single generator are examples of commutative, linearly ordered semigroups which are moreover Archimedean.

1. In [2], a condition weaker than diminishing orbital diameters was introduced for semigroups of selfmappings on a convex subset of a normed linear space.

Definition. Let X be a convex subset of a normed linear space and $\mathcal{F}: X \to X$ a semigroup of selfmappings. $\mathcal{F}$ is said to have convex diminishing orbital diameters if and only if $\forall x \varepsilon X$ with $0 < \delta(F(x)) < \infty$, $\exists p \varepsilon \overline{co} F(x)$ such that $\delta(F(p)) < \delta(F(x))$.

Among other things, the following theorem was proved in [2]:

Theorem. Let X be a non-empty, convex, weakly compact subset of a Banach space and $F: X \to X$ be a commutative, Archimedean, linearly ordered semigroup of nonexpansive mappings having convex diminishing orbital diameters. Suppose there is a $g \varepsilon F$ (with $g \neq Id$) such that g has diminishing orbital diameters, then F has a common fixed point.

As an immediate consequence, the following corollary is obtained.

Corollary. Let X be a non-empty, convex, weakly compact subset of a Banach space and $F: X \to X$ be a flow of nonexpansive mappings having convex diminishing orbital diameters. Suppose there is a $g \varepsilon F$ (with $g \neq Id$) such that g has diminishing orbital diameters, then F has a common fixed point.

2. In [3], the following definition for a semigroup of Lipschitz mappings is introduced.

Definition. A semigroup $F = \{f_\alpha : \alpha \varepsilon A\}$ of Lipschitz mappings of a Banach space is called eventually - nonexpansive if the family of Lipschitz constants $\{k_\alpha : \alpha \varepsilon A\}$ satisfies the following conditions: $\forall \varepsilon > 0$, $\exists \gamma \varepsilon A$ such that whenever $f_\beta \varepsilon f_\gamma F = \{f_\gamma f_\alpha : f_\alpha \varepsilon F\}$, then $k_\beta < 1 + \varepsilon$.

The following theorem, which is a generalization of the result obtained in [4], was proved in [3].

Theorem. If K is a non-empty, closed, convex and bounded subset of a uniformly convex Banach space and $F: K \to K$ is an eventually-non-expansive, commutative, linearly ordered semigroup of mappings, then F has a common fixed point.

For the proof of the above result, see [3].

References

[1] Horn, W.A., Some fixed point theorems for compact maps and flows in Banach spaces, Trans. Amer. Math. Soc. 149 (1970), 391-404.

[2] Kiang, M.T., Semigroups with diminishing orbital diameters, Pac. J. Math. 41, no.1, (1972) 143-151.

[3] Kiang, M.T., A fixed point theorem for eventually nonexpansive semigroups of mappings, J. Math. Analy. and App. (to appear).

[4] Goebel, K., and Kirk, W.A., A fixed point theorem for asymptotically-nonexpansive mappings, Proc. Amer. Math. Soc. 35, no. 1 (1972) 171-174.

Department of Mathematics
Saint Mary's University
Halifax, Nova Scotia

CARISTI'S FIXED POINT THEOREM AND THE THEORY OF NORMAL SOLVABILITY

W.A. Kirk[(1)]

Introduction

Let X and Y be Banach spaces and f a mapping of X into Y. The general problem we consider in this paper is that of determining the global implications of certain local assumptions on $f(X)$. We consider first mappings f of X into Y which are strongly ϕ-accretive in the sense that there exists $c > 0$ such that for all $x, u \in X$, $(f(x) - f(u), \phi(x-u)) \geq c||x - u||^2$, where ϕ is an appropriate mapping of X into Y^*. If this inequality is assumed to hold only for x sufficiently near u, then f is said to be locally strongly ϕ-accretive. In Section 1 we extend a surjectivity theorem of F. Browder for strongly ϕ-accretive mappings to the localized class. In doing so we circumvent the original normal solvability approach by an application of a fixed point theorem of J. Caristi in the space X. In Section 2 we combine this fixed point theorem with another approach of Browder to prove a mapping theorem (essentially Theorem 1 (I) of [7]) for a certain class of normally solvable mappings. For this result we apply Caristi's theorem in the space Y and relax assumptions on the space X. We conclude with a brief review of the theory of normal solvability.

1. Let X and Y be real Banach spaces with Y^* the conjugate space of Y, and let ϕ be a mapping of X into Y^* such that $\phi(X)$ is dense in Y^*, with

$$||\phi(x)||_{Y^*} = ||x|| \ , \ \phi(\xi x) = \xi\phi(x) \ ,$$

---

(1) Research supported in part by National Science Foundation grant GP 18045.

for all $x \in X$, $\xi \geq 0$. The mapping $f : X \to Y$ is said to be <u>strongly</u> ϕ-<u>accretive</u> if there exists $c > 0$ such that for all x , $u \in X$,

$$(1) \qquad (f(x) - f(u), \phi(x-u)) \geq c||x - u||^2 .$$

Mappings of this type have been studied by several authors in order to obtain simultaneously generalizations of mapping theorems for monotone mappings from X to X^* and for accretive mappings from X to X . Among the results of [6] is a surjectivity theorem for strongly ϕ-accretive nonlinear mappings $f : X \to Y$ obtained by application of a result in the general theory of normal solvability. This theory, introduced by Pohozhayev [13-15] and extensively developed by Browder in [1-7], is based on the premise that $f(X)$ is closed in Y , and it seeks to draw conclusions about the global structure of $f(X)$ from local assumptions (originally, differentiability assumptions) on f . Since (1) is easily seen to imply that $||f(x) - f(u)|| \geq c||x - u||$, x , $u \in X$, it follows that continuous strongly ϕ-accretive mappings are normally solvable and thus amenable to this approach.

The specific result of [6] to which we allude above (Theorem 4 - II) states that if $f : X \to Y$ is strongly ϕ-accretive, then appropriate assumptions on Y and Y^* along with the condition that for each $u \in X$

$$||x - u||^{-1/2} \, ||f(x) - f(u)|| \to 0 \quad \text{as } x \to u$$

imply that $f(X)$ is dense in Y and hence, that $f(X) = Y$. It is our object here to extend this result (retaining the original assumptions on Y , Y^*) to the more general class of mappings defined below. Continuous mappings of this class are not <u>a priori</u> normally solvable, but even so, because of the localized nature of our assumptions our main result (Theorem 4) highlights further the manner in which global conclusions can follow from local assumptions.

<u>Definition</u> 1. Suppose $\phi : X \to Y^*$ satisfies the assumptions stated at the outset. The mapping $f : X \to Y$ is said to be <u>locally strongly</u> ϕ-<u>accretive</u> if corresponding to each $y_0 \in Y$ and $r > 0$ there exists $c > 0$ such that the following condition holds.

(1.1) If $u \in X$ with $||f(u) - y_0|| \leq r$, then

$$(f(x) - f(u), \phi(x-u)) \geq c||x-u||^2$$

for all $x \in X$ sufficiently near u.

The present paper is a continuation of our approach of [11] in which we obtain a normal solvability result for mappings $f : X \to Y$ based upon an application in the space Y of a fundamental fixed point theorem due to Caristi [8, 9]. (An improved formulation of Caristi's result suggested by Felix Browder is actually used.) Here we apply this same theorem (Theorem 1 below) in the space X and thereby circumvent the normal solvability approach in deriving our result for strongly ϕ-accretive mappings.

Theorem 1. ([8, 9]). Let (M,d) be a complete metric space and $g : M \to M$ an arbitrary mapping. If there exists a lower semi-continuous function ϕ mapping M into the nonnegative real numbers such that for each $x \in M$,

$$d(x,g(x)) \leq \phi(x) - \phi(g(x)),$$

then g has a fixed point in M.

In the theorems below we use $B_r(y)$ to denote the closed ball centered at y with radius $r > 0$.

We derive our result for ϕ-accretive mappings, Theorem 4, from the following more basic results.

Theorem 2. Let X be a complete metric space, Y a normed linear space, and f a continuous mapping of X into Y. Suppose that for each $y_0 \in Y$ there exist constants $\alpha > 0$, $r > 0$ such that the following condition holds.

(2.1) Corresponding to each $x \in X$ with $f(x) \in B_r(y_0)$ there exists $\delta = \delta(x) > 0$ such that if $y \in B_\delta(f(x)) \cap f(X)$ then $d(x,v) \leq \alpha\, ||f(x) - y||$ for some $v \in f^{-1}(y)$.

Then, if $y_0 \in \operatorname{int} \overline{f(X)}$, $y_0 \in f(X)$. If $f(X)$ is dense in Y, then $f(X) = Y$.

Theorem 3. Let X be a complete metric space, Y a uniformly convex Banach space for which the duality mapping $J : Y \to Y^*$ is lipschitzian on bounded sets. Suppose $f : X \to Y$ is continuous and suppose that *for each* $y_0 \in Y$ *there exist constants* $\alpha > 0$, $r > 0$ *such that the following conditions hold*:

(3.1) $B_r(y_0) \cap f(X) \neq \phi$.

(3.2) *If* $y_0 \notin \overline{f(X)}$, *then for each* $y \in B_r(y_0) \cap f(X)$ *there exists* $v \in f(X)$ *such that* $(J(y_0 - y), v - y) \geq M||v - y||^2$, *where* M *is the Lipschitz constant of* J *on* $B_r(y_0)$.

(3.3) *Condition* (2.1) *of Theorem* 2.

Then $f(X) = Y$.

In Theorem 4 we assume that $f : X \to Y$ has a continuous local inverse on $f(X)$. By this we mean that if $y_j \to f(x)$ with $y_j \in f(X)$, then there exists $z_j \in f^{-1}(y_j)$ such that $z_j \to x$.

Theorem 4. *Let* X *and* Y *be Banach spaces,* f *a strongly* ϕ-*accretive mapping of* X *into* Y *in the localized sense of Definition* 1, *and suppose that* f *has a continuous local inverse on* $f(X)$. *Suppose* Y *and* Y^* *are uniformly convex with the duality mapping* $J : Y \to Y^*$ *satisfying a Lipschitz condition, suppose* $\phi(X) = Y^*$, *and suppose* f *satisfies the following* lip(1/2)-*condition*.

(4.1) *For each* $u \in X$,

$$||x - u||^{-1/2} \, ||f(x) - f(u)|| \to 0 \text{ as } ||x - u|| \to 0 .$$

Then $f(X) = Y$.

Proof of Theorem 2. Let $y_0 \in \operatorname{int} \overline{f(X)}$ and choose $\gamma \in (0,r)$ so that $B_\gamma(y_0)$ is contained in $\overline{f(X)}$. Suppose $f(x) \in B_r(y_0)$ and let $\eta > 1$ be fixed. Assume $f(x) \neq y_0$ and select $w \in \operatorname{seg}[f(x), y_0]$ so that $w \neq y_0$ and $0 < ||f(x) - w|| < \delta(x)$ where $\delta(x)$ is the constant of Condition (2.1). Since $w \in \overline{f(X)}$ there exists a sequence $\{y_j\}$ in $f(X)$ such that $y_j \to w$. Also since $0 < ||f(x) - w|| = ||f(x) - y_0|| - ||w - y_0||$ it follows that for fixed j sufficiently large,

$$(2) \qquad 0 < ||f(x) - y_j|| \leq \eta[||f(x) - y_0|| - ||y_j - y_0||] ,$$

and we may further assume $||f(x) - y_j|| \leq \delta(x)$. By (2.1) there exists $z \in f^{-1}(y_j)$ such that

$$(3) \qquad d(x,z) \leq \alpha||f(x) - y_j|| ,$$

where α is the constant of the theorem depending on y_0. Thus

(4) $$d(x,z) \leq \alpha\eta[||f(x) - y_0|| - ||y_j - y_0||] .$$

Because f is continuous the set

$$S = \{x \in X : ||f(x) - y_0|| \leq r\}$$

is closed, and since $||f(x) - y_0|| \leq r$, (2) implies $||y_j - y_0|| \leq r$; hence $z \in S$. Therefore we may define a mapping g of S into S by taking $g(x) = z$, with z determined as above if $f(x) \neq y_0$ and $z = x$ if $f(x) = y_0$. If $\phi : S \to R^+$ is defined by taking $\phi(x) = \alpha\eta||f(x) - y_0||$ then from (4), $d(x,g(x)) \leq \phi(x) - \phi(g(x))$, $x \in S$, so by Theorem 1 there exists $x \in S$ such that $g(x) = x$. This implies that $z = x$ and $y_j = f(x)$ for appropriate j contradicting (2) and implying $y_0 = f(x)$ for some x in X.

Proof of Theorem 3. The proof involves showing that the set Y_0 of points of Y which have a 'nearest' point in $\overline{f(X)}$ is contained in $\overline{f(X)}$ and then applying a theorem of Edelstein [10] which asserts that Y_0 is dense in Y. This can be accomplished by replacing $f(X)$ with $\overline{f(X)}$ in Browder's proof of Theroem 3 of [6]. With this and condition (3.3), Theorem 3 follows immediately from Theorem 2.

Proof of Theorem 4. We apply Theorem 3 and follow the proof of Theorem 4 of [6]. Let $y_0 \in Y$ be fixed, and let $y \in f(X)$. Choose $v \in X$ so that $||v|| = ||y - y_0||$, $\phi(v) = J(y - y_0)$. Suppose $r = ||y - y_0||$, and let $c > 0$ be the constant of Definition 1 corresponding to y_0, r. Suppose $f(u) = y$ and let $x_t = u + tv$. Then for $t > 0$ sufficiently small the condition of Definition 1 implies

$$t(f(x_t) - f(u), \phi(v)) \geq ct^2||y_0 - y||^2 ,$$

i.e.,

$$(f(x_t) - y, J(y_0-y)) \geq ct||y_0 - y||^2 .$$

On the other hand the lip(1/2)-condition on f implies

$$||f(x_t) - f(u)||^2 \leq \varepsilon(t)||x_t - u|| \leq \varepsilon(t)t||y_0 - y|| ,$$

where $\varepsilon(t) \to 0$ as $t \to 0^+$. Hence, given $M > 0$ it is possible to choose $t > 0$ so small that

$$M\|f(x_t) - f(u)\|^2 < (f(x_t) - f(u), J(y_0 - y)) \quad .$$

This establishes condition (3.2) of Theorem 3. Condition (3.3) also holds, for if $x \in X$ with $f(x) \in B_r(y_0)$ then by Definition 1,

$$(f(z) - f(x), \phi(z-x)) \geq c\|z - x\|^2$$

for all $z \in X$ sufficiently near x . Thus for such z

$$\begin{aligned} c\|z - x\|^2 &\leq (f(z) - f(x), \phi(z-x)) \\ &\leq \|f(z) - f(x)\| \, \|\phi(z - x)\| \\ &= \|f(z) - f(x)\| \, \|z - x\| , \end{aligned}$$

and since f has a continuous local inverse on $f(X)$, (3.3) holds for $\alpha = c^{-1}$. Since continuity of f follows from (4.1), all the assumptions of Theorem 3 are satisfied, completing the proof.

2. We now suppose X is a topological space, U a subset of X, $W = X - U$. Let f be a mapping of X into the Banach space Y with $f(X)$ a closed subset of Y . For each point y_0 of $f(U)$ with $r_0 = \text{dist}(y_0, f(W)) > 0$, assume that there exists $r > 0$ with $r < r_0$ such that the following condition holds.

(I) There exists $p < 1$ such that for each $x \in f^{-1}(B_{r/2}(y_0))$ and each $y \in B_{r/2}(f(x))$ there exists $u \in X$ and $\xi \geq 1$ such that

$$\|\xi(f(u) - f(x)) - (y - f(x))\| \leq p\|y - f(x)\| \quad .$$

Below we use Caristi's theorem to prove a version of Theorem 1 (I) of [7]. We show that if $f(X)$ is closed in Y , then the above basic assumption implies that f is <u>full over</u> U . This means that any component V of $Y - \overline{f(W)}$ which meets $f(U)$ must lie in $f(U)$. The connection between theorems of this type and continuation of solutions of functional equations under homotopy has been discussed previously by Browder (cf., [7, Theorem 2]).

<u>Theorem</u> 5. *Let X be a topological space, U a subset of X , $W = X - U$. Let f be a mapping of X into the Banach space Y with $f(X)$ closed in Y . If f satisfies condition (I), then f is full over Y .*

<u>Proof</u>. As observed in [7], it suffices to prove that the assumptions of the theorem imply the following assertion.

Assertion. For each point y_0 of $f(U) - \overline{f(W)}$, there exists a positive number δ such that the closed ball $B_\delta(y_0)$ lies in $f(U)$.

To prove the assertion, let y_0 be any point of $f(U) - \overline{f(W)}$. Then there exists $d > 0$ such that $B_d(y_0) \cap f(W) = \phi$, and we may suppose $d < r/2$ where r is the constant of Condition (I) associated with y_0. Let $d_1 = d/3$ and suppose $y \in B_{d_1}(y_0)$. We complete the proof by showing that $y \in f(X)$ and hence that $y \in f(U)$.

Suppose $y \notin f(X)$. Let $s = \text{dist}(y, f(X))$ and suppose $y_1 \in f(X)$ with $||y - y_1|| \leq 2s$. In particular, observe that $s \leq d_1$ and thus

$$||y_1 - y_0|| \leq ||y_1 - y|| + ||y - y_0|| \leq 2s + d_1 \leq d .$$

Hence $y_1 \in B_d(y_0)$. By Condition (I) there exist $u \in X$ and $\xi \geq 1$ such that

$$||\xi(f(u) - y_1) - (y - y_1)|| \leq p||y - y_1|| . \tag{5}$$

The proof of Theorem 2 of [11] shows that the above inequality implies

$$||y_1 - f(u)|| \leq [(1+p)/(1-p)][||y_1 - y|| - ||f(u) - y||] . \tag{6}$$

Also since $||y - y_1|| \leq 2s$, it follows from (6) that $||f(u) - y|| \leq 2s$. Thus if $B = \{z \in f(X) : ||z - y|| \leq 2s\}$ we may define a mapping g of B into B by taking $g(y_1) = f(u)$ with $f(u)$ as in (5). Defining $\phi : B \to R^+$ by $\phi(y_1) = [(1+p)/(1-p)]\, ||y_1 - y||$, we have ϕ continuous with (6) implying

$$||y_1 - g(y_1)|| \leq \phi(y_1) - \phi(g(y_1)) , \quad y_1 \in B .$$

Since $f(X)$ is closed, B is complete so by Theorem 1 $g(y^*) = y^*$ for some $y^* \in B$. By (5) this can happen only if $y = y^* \in f(X)$ contradicting our original assumption that $y \notin f(X)$.

3. In order to place the result of the previous section in appropriate perspective we briefly review the development of the theory of normal solvability, a theory which in the linear case dates from the work of Hausdorff in 1932.

Let X and Y be Banach spaces and $L : X \to Y$ a continuous linear mapping. If X^* and Y^* denote the respective conjugate spaces of X and Y and $L^* : Y^* \to X^*$ the adjoint mapping of L, then a necessary condition for solvability of the equation

(7) $$Lx = y$$

for given $y \in Y$ is that $y \in (N(L^*))^{\perp}$. (Here $N(L^*)$ denotes the nullspace of L^* and $(N(L^*))^{\perp}$ its annihilator in Y. Thus $N(L^*) = \{y^* \in Y^* : L^*y^* = 0\}$ and $(N(L^*))^{\perp} = \{y \in Y : \langle y^*,y\rangle = 0 ,$ $y^* \in N(L^*)\}$.)

The linear equation (7) is called <u>normally solvable</u> (Hausdorff) if the necessary condition $y \in (N(L^*))^{\perp}$ is also a sufficient condition for solvability (i.e., if it is the case that $\langle y^*,y\rangle = 0$ for all $y^* \in N(L^*)$ implies $Lx = y$ for some $x \in X$). In his survey [12] of solvability in the linear case, Kolomy notes that the proof of Banach's <u>closed range theorem</u> given in [17, p. 205] actually shows that a densely defined linear mapping $T : D \to Y$, $\bar{D} = X$, is normally solvable if and only if $T(D)$ is closed in Y.

We now describe the original results of Pohozhayev in the non-linear case. Suppose $f : X \to Y$ is continuous and has continuous Gateaux derivative df_x at each $x \in X$. If the equation

(8) $$f(x) = y$$

is solvable for given $y_0 \in Y$, i.e. if there exists $x \in X$ such that $f(x) = y_0$, then trivially $y_0 - f(x) \in (N(df_x^*))^{\perp}$ where df_x^* denotes the adjoint of the linear mapping df_x. Pohozhayev calls the equation (8) <u>normally solvable</u> if f is of class C^1 and if the following conditions hold.

1) For each $y_0 \in Y$ there exists $x_0 \in X$ such that $||f(x_0) - y_0|| = \text{dist}(y_0,f(x))$, and

2) for such x_0, $y_0 - f(x_0) \in (N(df_x^*))^{\perp} \Rightarrow y_0 = f(x_0)$.

For reflexive spaces this definition is equivalent to the one given in the linear case.

Theorem 1 of [14] gives sufficient conditions for normal solvability of the equation $y = f(x)$. It may be formulated as follows.

<u>Theorem</u> 6. <u>Let</u> X <u>and</u> Y <u>be Banach spaces spaces with</u> Y <u>reflexive, suppose</u> $f : X \to Y$ <u>is continuously differentiable with</u> $f(X)$ <u>weakly closed in</u> Y, <u>and suppose the following condition holds</u>.

(6.1) $\quad ||f(x) - y_0|| = \text{dist}(y_0, f(X)) \Rightarrow y_0 - f(x) \in (N(df_x^*))^{\perp}$.

Then $y_0 \in f(X)$.

From the above it follows that if Y is reflexive, $f : X \to Y$ continuously differentiable, and $f(X)$ weakly closed Y, then the assumption that $N(df_x^*) = \{0\}$ for all $x \in X$ implies $f(X) = Y$. Subsequently Pohozhayev proved the following result for mappings f with strongly closed range.

Theorem 7 ([15]). *Let X and Y be Banach spaces with Y uniformly convex, and suppose $f : X \to Y$ is Gateaux differentiable with $f(X)$ closed in Y. If $N(df_x^*) = \{0\}$ for each $x \in X$, then $f(X) = Y$.*

In proving the above theorem, Pohozhayev uses the result of Edelstein [10] that if S is a closed subset of a uniformly convex space Y, then the set

$$\{y \in Y : \text{dist}(y,S) \text{ is assumed in } S\}$$

is dense in Y. (In [16] Swaminathan observes that Pohozhayev's result holds for the wider class of all spaces Y having the above property.)

In a series of recent papers ([1-7]), Browder has considerably sharpened and generalized the above results. His basic result for arbitrary Banach spaces Y, Theorem 8 below, requires no differentiability (or continuity) assumption on f. When applied to Theorem 6, this result shows that if (6.1) is weakened to: $f^{-1}(B_r(y_0)) \neq \phi$ for some $r > 0$, and $x \in f^{-1}(B_r(y_0)) \Rightarrow y_0 - f(x) \in (N(df_x^*))^{\perp}$, then it follows that Y may be taken as an arbitrary Banach space and at the same time $f(X)$ need only be assumed closed rather than weakly closed. (The assumption that $f(X)$ is closed in Y is now taken as the defining premise in the study of normal solvability.) Browder's proof of Theorem 8 is intricate, based upon sharp results in the geometry of arbitrary Banach spaces.

Theorem 8 ([4,6]). *Let X be a topological space, Y a Banach space, and f a mapping of X into Y with $f(X)$ closed in Y. Suppose for a given point $y_0 \in Y$ the following condition holds.*

(8.1) *There exist constants $r > 0$, $p < 1$ such that*

(a) $B_r(y_0) \cap f(X) \neq \phi$, *and*

(b) *For each $y \in B_r(y_0) \cap f(X)$ there exists a sequence $\{y_j\}$ in $f(X)$, with $y_j \neq y$ for each j, such that $y_j \to Y$ and a* *sequence* $\{\xi_j\}$ *of nonnegative real numbers such that for each* j

$$||\xi_j(y_j-y) - (y_0-y)|| \leq p||y_0-y|| .$$

Then $y_0 \in f(X)$.

The following definition (from which assumption (b) of the preceding theorem evolved) and Proposition 1 below provide the connection between Browder's assumptions on f in Theorem 8 and Pohozhayev's differentiability hypotheses.

Definition ([3]). Let X be a topological vector space, f a mapping of X into the Banach space Y, and $x \in X$. The *asymptotic direction set* of f at x is the set

$$D_x(f) = \bigcap_{\varepsilon>0} \overline{D_{x,\varepsilon}(f)}$$

where

$$D_{x,\varepsilon}(f) = \{y\varepsilon Y : y = \xi(f(u)-f(x)),\ u\varepsilon X,\ \xi\geq 0,\ ||f(u)-f(x)|| < \varepsilon\} .$$

Proposition 1. ([3]). *Let* X *be a topological vector space,* Y *a Banach space, and* f *a mapping of* X *into* Y *which is Gateaux differentiable at* $x \in X$ *with continuous derivative* df_x. *Let* $df_x^* : Y^* \to X^*$ *denote the conjugate mapping of* df_x. *Then*

$$(N(df_x^*))^{\perp} = \overline{df_x(X)} \subset D_x(f) .$$

Condition (8.1) is easily seen to follow from the following assumption (introduced in [3]).

(8.1)´ There exists $r > 0$, $p < 1$ such that

(a´) $B_r(y_0) \cap f(X) \neq \phi$, and

(b´) if $f(x) \in B_r(y_0)$, then

$$\text{dist}(y_0-f(x), D_x(f)) \leq p||y_0-f(x)|| .$$

If the assumptions of Theorem 6 hold then by (6.1), $y_0 - f(x) = \text{dist}(y_0, f(X))$ implies $\text{dist}(y_0-f(x), (N(df_x^*))^{\perp}) = 0$, and thus by Proposition 1 $\text{dist}(y_0-f(x), D_x(f)) = 0$. Therefore (under these assumptions) condition (6.1) implies condition (8.1)´ (with $r = \text{dist}(y_0,f(X))$), and this in turn implies (8.1). It follows that Pohozhayev's result corresponds to a special case of Browder's Theorem 8 in which Y is assumed reflexive, $f(X)$ is weakly closed in Y, and (8.1) holds for all $p < 1$. Theorem 7 corresponds to a global version of Theorem 8 (with the stated assumption (8.1) holding for all $y \in Y$) in which Y is uniformly convex and $p < 1$ arbitrary.

In [11] Caristi's theorem is used in the same manner as in Section 2 to give a quick proof of a slightly sharpened version of Theorem 8 (with (8.1) weakened to an assumption similar to condition (I)).

References

[1] Browder, F.E., On the Fredholm alternative for nonlinear operators. Bull. Amer. Math. Soc., 76(1970), 993-998.

[2] Browder, F.E., Normal solvability and the Fredholm alternative for mappings in infinite dimensional manifolds, J. Functional Analysis 8(1971), 250-274.

[3] Browder, F.E., Normal solvability for nonlinear mappings in Banach spaces, Bull. Amer. Math. Soc. 77 (1971), 73-77.

[4] Browder, F.E., Normal solvability for nonlinear mappings and the geometry of Banach spaces, "Problems in Nonlinear Analysis", pp. 37-66, Edizioni Cremonese, Roma, Italy, 1971.

[5] Browder, F.E., Normal solvability and existence theorems for nonlinear mappings in Banach spaces, "Problems in Nonlinear Analysis", pp. 19-35, Edizioni Cremonese, Roma, Italy, 1971.

[6] Browder, F.E., Normal solvability and ϕ-accretive mappings of Banach spaces, Bull. Amer. Math. Soc. 78 (1972), 186-192.

[7] Browder, F.E., Normally solvable nonlinear mappings in Banach spaces and their homotopies, J. Functional Analysis, 17 (1974), 441-446.

[8] Caristi, J., The fixed point theory for mappings satisfying inwardness conditions, Ph.D. Thesis, University of Iowa, 1975.

[9] Caristi, J., Fixed point theorems for mappings satisfying inwardness conditions. Trans. Amer. Math. Soc. (to appear).

[10] Edelstein, M., On nearest points of sets in uniformly convex Banach spaces. J. London Math. Soc. 43 (1968), 375-377.

[11] Kirk, W.A., and Caristi, J., Mapping theorems in metric and Banach spaces, Bull de l'Academie Polonais des Sciences (to appear).

[12] Kolomy, J., Normal solvability and solvability of nonlinear equations, "Theory of Nonlinear Operators: Proceedings of a Summer School Oct. 1972 (Neuendorf)", Akademie-Verlag, Berlin, 1974.

[13] Pohozhayev, S.I., On the normal solvability of nonlinear operators, Dokl. Akad. Nauk SSSR 184 (1969), 40-43.

[14] Pohozhayev, S.I., On nonlinear operators having weakly closed images and quasilinear elliptic equations, Mat. Sb. 78(1969), 237-259.

[15] Pohozhayev, S.I. Normal solvability and nonlinear mappings in uniformly convex Banach spaces, Functional Anal. Appl. 3(1969) 80-84.

[16] Swaminathan, S., On the closed range theorem for nonlinear operators, Proc. of the Third Prague Topol. Sympos. (1971), Academia Prague (1972), 417-418.

[17] Yosida, K., Functional Analysis. Springer-Verlag, Berlin, 1965.

SOME FIXED POINT THEOREMS AND THEIR APPLICATIONS TO W*-ALGEBRAS

Anthony To-Ming Lau

1. Introduction

In this talk, we shall discuss several common fixed point properties for semigroups of mappings on compact (or weakly compact) convex subsets of a separated locally convex space, their relations, and some open problems concerning them. We shall also give at the end two applications of the fixed point properties to the theory of W*-algebras.

I would like to thank the Department of Mathematics, Dalhousie University, for their kind invitation to speak and participate in the seminar on fixed point theory and its applications.

2. Preliminaries

Let S be a semigroup, and let $\ell_\infty(S)$ be the space of bounded real-valued functions on S with the supremum norm. For each $a \in S$, let ℓ_a and r_a be the left and right translation operators defined by $(\ell_a f)(s) = f(as)$, $(r_a f)(s) = f(sa)$ for all $f \in \ell_\infty(S)$, $s \in S$. If X is a closed linear subspace of $\ell_\infty(S)$ invariant under the translation operators and contains the constant functions, then a continuous linear functional ϕ on X is called a <u>left invariant mean</u> on X if $||\phi|| = \phi(1) = 1$ and $\phi(\ell_a f) = \phi(f)$ for all $a \in S$, $f \in X$. The semigroup S is called <u>left amenable</u> if $\ell_\infty(S)$ has a left invariant mean. As well known, any commuting semigroup is left amenable, and any solvable group is left amenable. However, the free group on two generators is not left amenable. For more information on amenable semigroups, we refer our readers to the survey articles of Day [2], [4] and Greenleaf's book [6].

A semigroup S is called <u>left reversible</u> if the intersection of any two (hence finitely many) right ideals is non-empty. The class of left reversible semigroups clearly includes all groups, all commuting semigroups, and more generally all left amenable semigroups.

A function f in $\ell_\infty(S)$ is <u>strongly almost periodic</u> (resp. <u>weakly almost periodic</u>) if $LO(f) = \{\ell_a f;\ a \in S\}$ is relatively compact in the sup norm (resp. weak) topology of $\ell_\infty(S)$. We shall denote by AP(s) (resp. WAP(s)) the closed subspace of $\ell_\infty(S)$ consisting of all strongly (resp. weakly) almost periodic functions.

If $\mathcal{S}$ is a semigroup of mappings on a subset X of a separated locally convex space (E,τ) into X, we say that $\mathcal{S}$ is <u>τ-equicontinuous</u> on X if for each τ-neighbourhood U of the origin, there exist a τ-neighbourhood of the origin V such that whenever $x-y \in V$, $x,y \in X$, we have $T(x) - T(y) \in U$ for all $T \in \mathcal{S}$; $\mathcal{S}$ is <u>τ-distal</u> if for any distinct elements x,y in X, 0 does not belong to the τ-closure of $\{T(x)-T(y);\ T \in \mathcal{S}\}$.

3. <u>Fixed point properties for semigroups of affine mappings</u>.

In [16], Ryll-Nardzewski proved the following important theorem:

<u>Theorem</u> (Ryll-Nardzewski [16]): Given any semigroup S, if $\mathcal{S} = \{T_s;\ s \in S\}$ is a representation of S as affine continuous self maps on a weakly compact convex set X of a separated locally convex space (E,τ), and $\mathcal{S}$ is τ-distal on X, then X contains a common fixed point for $\mathcal{S}$.

Namioka and Asplund [15] gave an elegant proof of Ryll-Nardzewski's theorem. In the case that X is assumed to be τ-compact, there is short and completely different proof due to F. Hahn [7].

In this section, we consider on the semigroup S the following fixed point properties:

(F_1): Whenever $\mathcal{S}$ is a representation of S as continuous affine self-maps on a compact convex subset X of a separated locally convex space, then X contains a common fixed point for $\mathcal{S}$.

(F_2): Whenever $\mathcal{S}$ is a representation of S as τ-equicontinuous affine self-maps on a weakly compact convex subset X of a separated locally convex space (E,τ), then X contains a common fixed point for $\mathcal{S}$.

(F_3): Whenever $\mathcal{S}$ is a representation of S as τ-equicontinuous self-maps on a τ-compact convex set X of a separated locally convex space (E,τ), then X has a common fixed point for $\mathcal{S}$.

Markov proved in [14] that any commuting semigroup has property (F_1) . Finally Day [3] proved:

Theorem (Day [3]): A semigroup S has property (F_1) if and only if S is left amenable.

If S is a semigroup with an identity and has no proper right ideals, then S has property (F_2). Indeed, let x,y be distinct elements in X . Choose V a τ-neighbourhood of the origin such that $x-y \notin V$. By equicontinuity, we can find a τ-neighbourhood U of the origin such that $T_s(u)-T_s(v) \in V$ for all $s \in S$ if $u-v \in U$, $u,v \in X$. It follows easily that the intersection of U with $\{T_s(x)-T_s(y);\ s \in S\}$ is empty. Consequently $\mathcal{S} = \{T_s\ ;\ s \in S\}$ is τ-distal on X . By Ryll-Nardzewski's theorem, X has a common fixed point for $\mathcal{S}$.

Theorem: A semigroup S has fixed point property (F_2) if and only if WAP(S) has a left invariant mean.

Proof: Assume that WAP(S) has a left invariant mean. Let Φ denote the convex hull of $\mathcal{S}$ in the product space E^X and $\bar{\Phi}$, $\bar{\mathcal{S}}$ denote the closure of Φ , $\mathcal{S}$ in $(E, \text{weak})^X$. Since Φ is convex, $\bar{\Phi}$ is also the closure of Φ in the product space $(E,\tau)^X$. By τ-equicontinuity of $\mathcal{S}$, $\bar{\Phi}$ consists entirely of continuous affine maps from (X,τ) into (X,τ) . Consequently, maps in $\bar{\Phi}$ are also continuous from (X, weak) into (X, weak). Since WAP(S) has a left invariant mean, it follows from [5, Lemma 5.2] that $C(\bar{\mathcal{S}})$ also has a left invariant mean. Consequently $\bar{\mathcal{S}}$ contains a compact topological group G which is also a left ideal (see [5]). Let $x \in X$, and $\hat{x}: G \to X$ be defined by $\hat{x}(g) = g(x)$. Let λ be the unique normalized Haar measure on G . Then the vector-valued integral $\int_G \hat{x}\,d\lambda$ defines an element x_0 in X which is fixed under G . It is easy to check that x_0 is a common fixed point for $\mathcal{S}$.

Conversely, if S has fixed point property (F_2), let X denote the set of means on WAP(S) i.e. all $\phi \in WAP(S)^*$ such that $||\phi|| = \phi(1) = 1$. Let τ be the topology on WAP(S) defined by the seminorms $\{p_f\ ;\ f \in WAP(S)\}$, where

$$p_f(\phi) = \sup\{|\phi(\ell_s f)|, |\phi(f)|;\ s \in S\} .$$

Then by Mackey Aren's theorem, the dual of WAP(S)* with respect to τ

is still WAP(S) and the representation $\{\ell_s^* ; s \in S\}$ on (X,τ) is clearly τ-equicontinuous and affine (in fact non-expansive with respect to $\{p_f ; f \in WAP(S)\}$). Any common fixed point in X is a left invariant mean on WAP(S) .

Kakutani [8] proved that if S is a group, then S has property (F_3). Hahn [7] pointed out that Kakutani's theorem can actually be obtained from his theorem. Sneperman [20], proved by modifying Kakutani's argument, that any left reversible semigroup has property (F_3). Sneperman's result can also be obtained by the following argument: if S is left reversible, and $\mathcal{S}$ is a representation of S as equicontinuous affine maps on a compact convex set X, let $\bar{\mathcal{S}}$ denote the closure of $\mathcal{S}$ in the product space X^X. Then $\bar{\mathcal{S}}$ is a compact topological semigroup with jointly continuous multiplication, and $\bar{\mathcal{S}}$ also consists of affine continuous self maps on X. Furthermore, since S is left reversible, $\bar{\mathcal{S}}$ is also left reversible. Hence $\bar{\mathcal{S}}$ must contain a compact group G which is a left ideal (see [5]). Pick $x \in X$, then exactly as in the proof of the last theorem, $\int_G \hat{x} d\lambda$ is a common fixed point for $\mathcal{S}$.

We prove in [11] the following:

Theorem (Lau [11]): A semigroup S has property (F_3) if and only if AP(S) has a left invariant mean.

4. Some open problems.

In this section, we shall state several problems relating the fixed point propertis discussed in section 3. The following diagram indicates the relationship among properties F_1, F_2, F_3 and left reversibility.

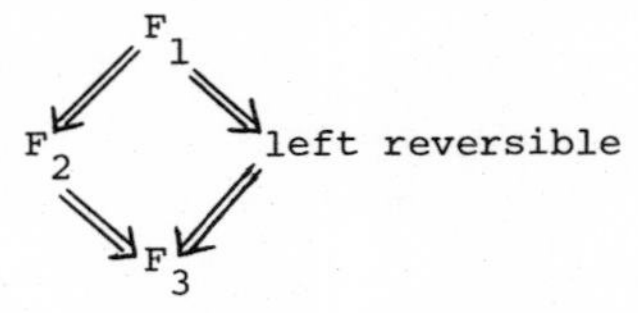

Since the free group on two generators fails to have property (F_1), it is clear that neither property F_2 nor left reversibility imply F_1. However the following problems seem to be open:

Problem 1: Does (F_3) imply (F_2) ?

Problem 2: Does left reversibility imply (F_2)?

Problem 3: Does (F_2) imply left reversibility?

If $\mathcal{S}$ is a semigroup of mappings on a subset X of a separated locally convex space (E,Q), where Q is a family of semi-norms determining the topology of E, we say that $\mathcal{S}$ is Q-non-expansive on X if $p(Tx-Ty) \leq p(x-y)$ for all $T \in \mathcal{S}$, $p \in Q$ and all x,y in X.

Consider on a semigroup S the following fixed point properties:

(G_1): Whenever $\mathcal{S}$ is a representation of S as weak*-continuous self-maps on a weak*-compact convex subset X of a dual Banach space $(E,||\cdot||)$, and $\mathcal{S}$ is $||\cdot||$-non-expansive on X, then X has a common fixed point for $\mathcal{S}$.

(G_2): Whenever $\mathcal{S}$ is a representation of S as weakly continuous, Q-non-expansive mappings on a weakly compact convex set X of a separated locally convex space (E,Q), then X contains a common fixed point for $\mathcal{S}$.

(G_3): Whenever $\mathcal{S}$ is a representation of S as Q-non-expansive mappings on a compact convex subset X of a separated locally convex space (E,Q), then X contains a common fixed point for $\mathcal{S}$.

R. DeMarr [23] proved that any commuting semigroup S has fixed point property (G_3) (DeMarr [23], deals only with Banach spaces, however his argument can be modified trivially to locally convex spaces). W. Takahashi [22] proved that any left amenable semigroup S has property (G_3). T. Mitchell [13] generalized both DeMarr and Takahashi's result by showing that any left reversible semigroup S has property (G_3). Finally we proved in [11] a semigroup S has property (F_3) if and only if S has property (G_3).

It is easy to see that if a semigroup S has property (G_1), then S has property (F_1). Indeed, if S has property (G_1), let X be the set of means on $\ell_\infty(S)$. Then $\{\ell_s^* ; s\in S\}$ is a representation of S as weak*-continuous, norm non-expansive mappings on the weak*-compact convex set X. Any common fixed point in X is a left invariant mean on $\ell_\infty(S)$. By Day's fixed point theorem, S has property (F_1). Similarly we have property (G_2) implies property (F_2).

Problem 4: Does property (F_1) imply property (G_1)?

Problem 5: Does property (F_2) imply property (G_2)?

In fact, even the following simpler problems seem to be unknown:

Problem 6: Does any commuting semigroup S have property (G_1)?

Problem 7: Does any commuting semigroup S have property (G_2)?

Problem 8: Does any group S have property (G_2)?

Partial solutions to problems 6, 7, 8 are known when the underlying space is either strictly convex, or when the set X has normal structure. For complete details, we refer the readers to [1], [12] and [9].

5. Applications to W*-algebras.

In this section, we shall briefly indicate two applications of the fixed point properties discussed in section 3 to the theory of W*-algebras. For general reference to W*-algebras, we refer the readers to [17].

A. W*-algebras with property P.

Let M be a W*-algebra acting on a Hilbert space H. Let M^u denote the group of unitary elements in M, and let M' denote the commutant of M. For each T in $\mathcal{B}(H)$, the bounded linear operators on H, let K_T denote the σ-closed convex hull of the set $\{v^*Tv;\ v\in M^u\}$, where σ denotes the ultra-weak topology on $\mathcal{B}(H)$. M is said to have property P if for each T in $\mathcal{B}(H)$, the sets K_T and M' have non-empty intersection. Property P was first introduced by Schwartz in [18].

Theorem: Let G be a group, and let $\mathcal{G} = \{U_g\ ;\ g\in G\}$ be a representation of G as unitary operators on a Hilbert space H. If G is amenable, then the W*-algebra generated by $\mathcal{G}$ has property P.

Proof: Let M denote the W*-algebra generated by $\mathcal{G}$. Let $T \in \mathcal{B}(H)$. For each $g \in G$, let $\tau_g: K_T \to K_T$ be defined by $\tau_g(V) = U_g^*VU_g$. Then each τ_g is affine, and σ-continuous. Since K_T is σ-compact and convex. An application of Day's fixed point theorem shows that there exist $V \in K_T$ such that $U_g^*VU_g = V$ for all $g \in G$. Clearly $V \in M'$.

B. Finite W*-algebras.

Let M be a W*-algebra, and $\mathcal{G}$ denote a group of $*$-automorphisms on M. M is said to be $\mathcal{G}$-finite if for each non-zero positive element x in M, there exist a non-zero $\mathcal{G}$-invariant normal state ϕ on

M such that $\phi(x) > 0$ (see [10]). In case G is the group of inner automorphisms, then this is precisely the class of finite W*-algebras in the usual sense.

Let $F(M,G)$ denote the σ-closed (σ denotes the ultraweak topology) subspace of M consisting of all x in M such that $g(x) = x$ for all $g \in G$. For each x in M, let C_x denote the σ-closed convex hull of $\{g(x); g \in G\}$.

The following theorem is due to Kovacs and Szücs [10]. We give here an alternate proof using the Ryll-Nardzewski fixed point theorem.

<u>Theorem</u> (Kovacs and Szücs [10]): If M is G-finite, then for each $x \in M$, the set $C_x \cap F(M,G)$ consists of exactly one element.

<u>Proof</u>: If M is G-finite, it follows from the theorem in [19, p. 255] that for each $\phi \in M_*$, where M_* is the unique predual of M, $\{\phi \circ g ; g \in G\}$ is relatively compact in the weak topology of M_*. Let τ be the topology on M determined by the family of seminorms $\{p_\phi ; \phi \in M_*\}$, where

$$p_\phi(x) = \sup\{|\phi(g(x))|; g \in G\} .$$

It follows from the Mackey Arens theorem that τ is a topology of the dual pair (M,M_*). An application of fixed point property (F_2) (which holds for all groups by Ryll-Nardzewski fixed point theorem) on the σ-compact convex set C_x shows that $C_x \cap F(M,G)$ is non-empty.

To see that $C_x \cap F(M,G)$ consists of exactly one element, it suffices to show that the G-invariant functionals in M_* separate elements in $F(M,G)$. Indeed, if $y \in F(M,G)$, and $y \neq 0$, pick $\gamma \in M_*$ such that $\gamma(y) \neq 0$. Another application of the Ryll-Nardzewski fixed point theorem on B_γ, the closed convex hull of $\{\gamma \circ g ; g \in G\}$ in M_*, shows that there exists $\gamma_0 \in B_\gamma$ such that $\gamma_0(y) = \gamma(y) \neq 0$.

REFERENCES

[1] Brodskii, M.S., D.P. Mil'man, "On the center of a convex set" (Russian), Dikl, Akad. Nauk SSSR <u>59</u> (1948), 837-840.

[2] Day, M.M., "Amenable semigroups", Illinois J. Math. <u>1</u>, pp. 509-544.

[3] Day, M.M., "Fixed point theorems for compact convex sets", Illinois J. Math. <u>5</u> (1961), pp. 585-589.

[4] Day, M.M., "Semigroups and amenability", 'Semigroups', Proc. of a Symposium at Wayne State University, Ed. K. Folley, New York, London, Academic Press, pp. 5-53.

[5] Deleeuw, K., and I. Glicksberg, "Applications of almost periodic compactifications", Acta Math. 105 (1961), pp. 63-97.

[6] Greenleaf, F., "Invariant means on topological groups", Van Nostrand, 1969.

[7] Hahn, F., "A fixed point theorem", Math. System Theory I (1967), pp. 55-57.

[8] Kakutani, S., "Two fixed point theorems concerning bicompact convex sets", Proc. Imp. Acad. Tokyo 14, pp. 242-245 (1938).

[9] Kiang, M.T., "Fixed point theorems for certain classes of semigroups of mappings", Trans. A.M.A. 189 (1974).

[10] Kovacs, I., and J. Szücs, "Ergotic type theorems in von Neumann algebras", Acta. Sci. Math. 27(1966), pp. 233-246.

[11] Lau, A.T., "Invariant means on almost periodic functions and fixed point properties", Rocky Mountain J. Math. 3 (1973), pp. 69-76.

[12] Lim, T.C., "Characterizations of normal structure", Proc. A.M.S. 43 (1974), pp. 313-319.

[13] Mitchell, T., "Fixed point of reversible semigroups of non-expansive mappings", Kodai Math. Seminar Rep. 22(1970), pp. 322-323.

[14] Markov, A.A., "Quelques théorèms sur les ensembles akéliens C.R. (Doklady) Acad. Sci. URSS 1, 311-313, (1936).

[15] Namioka, I., and E. Asplund, "A geometric proof of Ryll-Nardzewski's fixed point theorem", Bull. A.M.S. 73, 443-445 (1967).

[16] Ryll-Nardzewski, C., "On fixed points of semigroups of endomorphisms of linear spaces", Proc. Fifth Berkeley Symp. on Math. Stat. and Prob., Vol. 2, part I, 55-61. Berkeley-Los Angeles: Univ. of California Press (1967).

[17] Sakai, S., "C*-algebras and W*-algebras", Springer-Verlag (1971).

[18] Schwartz, J., "Two finite, non-hyperfinite, non-isomorphic factors", Comm. Pure Appl. Math. 16, 19-26 (1963).

[19] Stormer, E., "Invariant states of von-Neuman algebras", Math. Scand. 30 (1972), pp. 253-256.

[20] Sneperman, L.B., "A fixed point of semigroup of transformations and invariant integration on a bicompact semigroup", Vesci Akad. Navuk BSSR Ser Fix - Mat Navuk 1966 no. 4, 30-36 (Russian).

[21] Sneperman, L.B., "The fixed point of a semigroup of transformations and invariant integration", Interuniv. Sci. Symposium, General Algebra, Tartu Gos. Univ. Tartu, 1966, pp. 215-216 (Russian).

[22] Takahashi, W., "Fixed point theorem for amenable semigroups of non-expansive mappings", Kodai Math. Sem. Rep. 21 (1969), pp. 383-386.

[23] DeMarr, R.E., "Common fixed points for commuting contraction mappings", Pacific J. Math. 13 (1963), pp. 1139-1141.

Department of Mathematics
University of Alberta
Edmonton, Alberta

COMPUTING THE BROUWER FIXED POINT BY FOLLOWING THE CONTINUATION CURVE

Tien-Yien Li*

§1 Introduction

Let D be a bounded convex domain in R^n, and let $F: \bar{D} \to \bar{D}$ be continuous. The proof for the existence of x^0, such that $F(x^0) = x^0$, has been established for years. A numerical method which can be implemented by computer was first found by Scarf in 1967 [8]. This method, which is based on the use of Sperner's Lemma, has been extended in a number of ways. (See [1], [2]).

Recently, Kellogg, Li and Yorke ([4], [5]) has introduced a different method for computing a fixed point of a C^2 function F. The algorithm is modeled on a different proof of the Brouwer Theorem, given in [3, 6]. In this proof, one reduces the existence of a fixed point to the non-existence of a retraction H of $\bar{D}$ onto ∂D. The retraction is shown not to exist by assuming that it does and studying the curves $H^{-1}(x^0)$, $x^0 \varepsilon \partial D$. One then shows that the curves $H^{-1}(x^0)$ lead from $x^0 \varepsilon \partial D$ to the fixed point. Nonrigorous methods for following these curves can be useful in practice on 100 dimensional problems, though theoretically, difficulties could occur.

While Hirsch's proof [3] is obviously ideal for computer implementation, this fact seems to have been completely missed prior to [4]. More recently S. Smale and M. Hirsch have been experimenting with a number of schemes for computer implementation of their idea of following curves to a fixed point or to a zero of a map. They have recently informed us that they have experimented with implementing scheme quite similar to the one presented here. Their schemes are usually designed for problems concerning smooth proper maps $F: R^n \to R^n$ and are directed toward finding zeros of F by following curves on which $F(x)/|F(x)|$

is constant. While their objective has been to experiment with effective schemes and also to apply these ideas to the theory of economics, our objective in this paper has been to show that their type of scheme can be made rigorous when F'' is bounded. We show that it is possible to choose the 'step size' so that convergence to a fixed point is permitted.

We first describe the basic method in detail in section 2 and state in section 3 a theorem which gives the rigorous method for following the curves.

§2 The Basic Method

Let D be a convex bounded open subset of R^n, and let $F:\bar{D} \to D$ be continuous. We assume that F is twice continuously differentiable in D, and that the first and second derivatives of F have continuous extensions to $\bar{D}$. Let $C = \{x \varepsilon D: F(x) = x\}$ be the set of fixed points of F. We define a map $H: \bar{D}/C \to \partial D$ as follows. For $x \varepsilon \bar{D}/C$, we denote by $L(x)$ the ray from $F(x)$ passing through x, and we let $H(x)$ denote the intersection of this ray with ∂D. The main theorem, on which the algorithm is based, is as follows. Let $U \subset \bar{D}/C$ be the set of x for which $H'(x)$ has rank $n-1$. Notice that U is open in the relative topology of $\bar{D}$.

Theorem 1. For almost every $x^0 \in \partial D$, $H^{-1}(x^0) \subset U$. Also, the connected component of the set $H^{-1}(x^0)$ which contains x^0 is a smooth curve which may be parametrized by arc length and may be written $x(t)$, $0 \le t < T \le \infty$ where T is the arc length and $x(0) = x^0$. Furthermore $d(x(t),C) \to 0$ as $t \to T$.

The curve $x(t)$ given by Theorem 1 is called the *continuation curve starting at* x^0. It satisfies the equations

$$H(x(t)) = x^0 \qquad (1)$$

$$x(0) = x^0 .$$

In order to allow the curve, we see that $x(t)$ satisfies

$$\frac{d}{dt} x(t) = u(x(t)) \qquad 0 \le t < T < \infty \qquad (2)$$

where u is a null vector for $H'(x)$; that is

$$H'(x)u = 0$$

Based on (2), one may easily design a useful but nonrigorous iterative algorithm as follows. Namely, at $x \varepsilon X(t) \subset H^{-1}(x^0)$, we calculate $H'(x)$ and its 0-eigenvector u, and then use u as a tangent vector to follow along the curve $x(t)$.

Of course if u is a unit vector which is an eigenvector of $H'(x)$, then so is $-u$. Since we have been approximating a smooth curve only one of the pair u and $-u$ will be consistent with proceeding in an orderly way along the curve while the incorrect choice would result in a very abrupt reversal in direction. Throughout the remainder of this paper, we leave the distinction between u and $-u$ to the reader, for clarity of exposition. By this method, we no longer stay on the curve $x(t)$. We do, however, approximate the curve. Although this procedure allows rapid convergence to a fixed point in practice, it has some disadvantages theoretically. First of all, we integrate $x(t)$ over an indefinite interval. The error can theoretically build up and eventually leads to an unacceptable result. Secondly, because of the unestimated numerical error, one might jump from a "good" curve to a "bad" curve and the algorithm becomes invalid. A "bad" curve might be a closed loop contained in $H^{-1}(x^0)$. The description above is given in greater detail in [4, 5].

§3

In this section, we let D be the unit cube in R^n with one face lying on $x_1 = 0$ and $|F''| < M$, for some constant M where $F'' = \left(\frac{\partial^2 F_i}{\partial x_j \partial x_k} \right)$ and $|F''| = \max_{i,j,k} \frac{\partial^2 F_i}{\partial x_j \partial x_k}$. The following lemma is the well-known Newton-Kantorovich Theorem [7]:

Lemma 3.1. Assume that $G: R^n \to R^n$ is differentiable on a convex set D and that

$$|G'(x) - G'(y)| \leq \gamma |x - y|$$

for all $x, y \varepsilon D$. Suppose that there is a $y^0 \varepsilon D$ such that $||G'(y^0)^{-1}|| \leq \beta$, $||G'(y^0)^{-1} G y^0|| = \eta$, and $\alpha = \beta\gamma\eta < 1/2$. Then the Newton iterates

$$y^{k+1} = y^k - G'(y^k)^{-1} G(y^k) \quad k = 0,1,... \tag{3}$$

are well defined and converge to the solution y^* of $G(y) = 0$.

We now choose $x^0 \varepsilon \{x: x_1 = 0\} \subset \partial D$ such that if $x \varepsilon H^{-1}(x^0)$, $H'(x)$ is of rank $n - 1$ and let $x(t)$ be the curve with $H(x(t)) = x^0$ and $x(0) = x^0$. We say v is a null vector if $|v| = 1$ and $H'(x)v = 0$. The following result concerns an explicitly calculable radius of convergence γ.

Theorem 2. For a point x on the curve $x(\cdot)$, let v be a null vector of $H'(x)$. There exists a neighborhood N with radius $\gamma = \gamma(x, M, |x - F(x)|)$, such that the Newton iterates in (3), when starting from $y^0 = x + \gamma_1 v$, $\gamma_1 \leq \gamma$, for the equation

$$G(x) = 0$$

where $G \equiv (G_1, \ldots, G_n)$ is defined by

$$G_i(x) = H_i(x) - x_0 \quad \text{for } i = 2, \ldots, n$$

$$G_1(x) = (x - y^0) \cdot v$$

converges to a point x^* on the curve $x(\ )$.

If we choose t_1 the number such that $x^* = x(t_1)$, then we will have $t_1 > t_0$, and so x^* will be "further along the curve" than x^0.

Based on Theorem 2, we may design a new algorithm as indicated in the following figure:

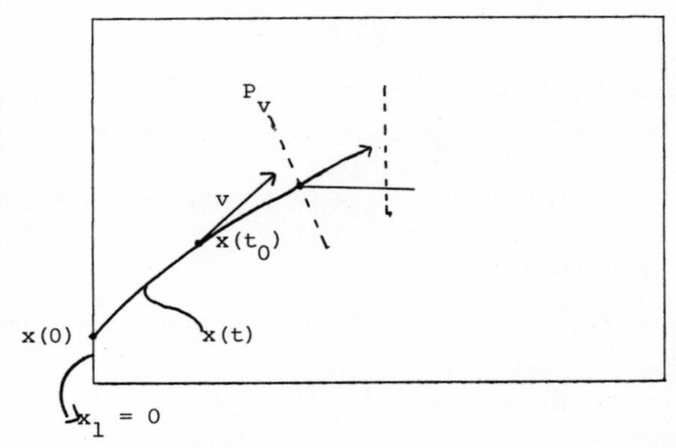

At any point $x(t_0)$ on the curve we first find the null vector v for $H'(x)$, calculate the number γ as described in Theorem 2, and remove along the tangent within the ball with radius γ and center $x(t_0)$. Then, we use Newton iteration on the set of equations $G(x) = 0$ as in Theorem 2. Namely, we find the point $x(t_1)$ of intersection of the hyperplane H_v with the curve $x(t)$. We then iterate this procedure to obtain $x(t_2), x(t_3), \ldots$ with $t_i < t_{i+1}$, and $t_i \to T$.

References

[1] Eaves, B.C., "Homotopies for computation of Fixed Points," Mathematical Programming, 3, 1-22.

[2] Eaves, B.C., and Saigal, R., "Homotopies for computation of fixed points on unbounded regions", Math Programming 3(1972), 225-237.

[3] Hirsch, M., "A proof of the nonretractibility of a cell onto its boundary", Proc. Amer. Math. Soc. 14(1963), pp. 364-365.

[4] Kellogg, R.B., Li, T.Y. and Yorke, J., "A method of continuation for calculating a Brouwer fixed point", to appear in Computing Fixed Points with Applications, S. Karamadian, ed., Academic Press, New York.

[5] Kellogg, R.B., Li, T.Y. and Yorke, J. "A constructive proof of the Brouwer Fixed Point Theorem and computational results" to appear in SIAM J. Numer. Anal. 13 (1976), 473-83

[6] Milnor, J.W., Topology from the Differentiable Viewpoint, University Press of Virginia, Charlottesville, 1965.

[7] Ortega, J.M., Numerical Analysis, a second course, Academic Press, New York.

[8] Scarf, H., "The approximation of fixed points of a continuous mapping, " SIAM J. Appl. Math. 15(1967), pp. 1328-1343.

Department of Mathematics
University of Utah
Salt Lake City, Utah 84112
U.S.A.

* Research partially supported by the National Science Foundation Grant MPS 74-24310.

SOME APPLICATIONS OF THE FIXED POINT INDEX IN ASYMPTOTIC FIXED POINT THEORY

Heinz-Otto Peitgen

Introduction

In [2] F.E. Browder, by rather elaborate arguments, proved the following asymptotic fixed point theorem:

Let C be a compact, convex, infinite dimensional subset of a Banach space and $f: C \to C$ continuous. Then f has a non-repulsive fixed point.

In [9] and independently in [11] it was observed that this theorem is a consequence of the fact that in this case a repulsive fixed point always has zero index. The result in [11] is stated for compact metric NAR-s and the proof makes essential use of what has recently been known as the 'mod p' theorem of Zabreiko-Krasnosel'skii [16] and Steinlein [15]. In [4] an index characterization for repulsive fixed points is proved for multivalued transformations acting on a compact metric ANR using only the 'mod p' theorem for the Lefschetz number [12] rather than the fixed point index. In this article we extend the methods used in [4] to the case of arbitrary metric ANR-s and give refined proofs for the characterization of repulsive and attractive fixed points. The consideration of arbitrary metric ANR-s is of interest, because this class of spaces contains important examples both in topology and analysis.

Preliminaries

In what follows an essential use will be made of the notion of the Lefschetz number in the generalized sense as given by J. Leray in [7] and the fixed point index for metric ANR-s developed by A. Granas in [5]. In particular we will use subsequently (cf. [5] lemma 2.1, 3.1) the following observation:

(i) Assume that in the category of graded vector spaces (over the rational numbers) and endomorphisms of degree zero the following diagram commutes

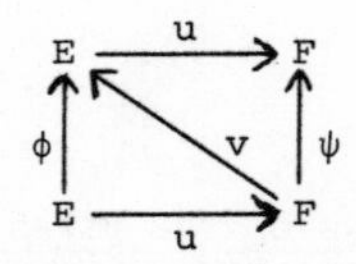

Then if ϕ or ψ is a Leray endomorphism (i.e. $\tilde{E} = E/N(\phi)$ is of finite type, where $N(\phi_q) = \bigcup_{n>0} (\ker \phi_q^n)$ then so is the other and in that case $\Lambda(\phi) = \Lambda(\psi)$. ($\Lambda(\phi) = \Sigma_q(-1)^q \operatorname{Tr}(\phi_q)$ is the generalized Lefschetz number and $\operatorname{Tr}(\phi_q) = \operatorname{trace}(\tilde{\phi}_q)$, $\tilde{\phi}: \tilde{E} \to \tilde{E}$, is the generalized trace in the sense of Leray [7]. The classical Lefschetz number, i.e. E is of finite type, is denoted by $\lambda(\phi)$.)

This property is a consequence of the fact $\operatorname{trace}(uv) = \operatorname{trace}(vu)$. H_* denotes the singular homology functor with rational coefficients and f_* is the abbreviation for $H_*(f)$.

(ii) Assume that in the category of topological spaces the following diagram commutes

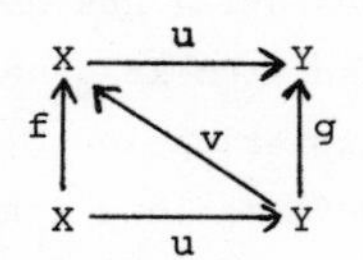

Then if one of the mappings f or g is a Lefschetz map (i.e. f_* or g_* is a Leray endomorphism) then so is the other and

$$\Lambda(f,X) = \Lambda(f_*) = \Lambda(g_*) = \Lambda(g,Y) .$$

Property (ii) follows from (i) by applying H_* .
For triples (X,f,U) , where X is a metric ANR, U is open in X and $f: U \to X$ is a compact 'admissible' map (i.e. $\operatorname{Fix}(f) = \{x \in U: f(x) = x\} \subset U$ is compact) a fixed point index $i(X,f,U)$ is defined in [5], which satisfies the standard properties including the strong normalization property $i(X,f,X) = \Lambda(f,X)$.

The class of metric ANR-s contains for example:
- convex subsets of linear normed spaces
- open subsets of linear normed spaces
- finite polyhedra

- compact manifolds
- Banach manifolds

All mappings are assumed to be continuous. If X is a topological space, $f: X \to X$ a map then f is said to be compact if $f(X) \subset K$ and $K \subset X$ is compact. If $x_0 \in X$ then $U(x_0)$ denotes the system of open neighborhoods of x_0 in X.

The notion of a repulsive fixed point is due to F.E. Browder [2] and is in some sense 'dual' to the notion of an attractive fixed point. In view of applications (cf. [9], [10]) we define repulsivity in a slightly more general situation, i.e. the mapping might not be defined continuously in the considered point.

Definition. Let X be a topological space and $x_0 \in X$.

(i) If $f: X\setminus\{x_0\} \to X$ is a map, x_0 is said to be repulsive for f relative $U \in U(x_0)$ if for all $V \in U(x_0)$ there is $n_0 \in \mathbb{N}$ such that $f^n(X\setminus V)$ is defined and

$$f^n(X\setminus V) \subset X\setminus U \text{ , whenever } n \geq n_0 \text{ .}$$

(ii) If $f: X \to X$ is a map, x_0 is said to be attractive for f relative $U \in U(x_0)$ if for all $V \in U(x_0)$ there is $n_0 \in \mathbb{N}$ such that

$$f^n(U) \subset V \text{ , whenever } n \geq n_0 \text{ .}$$

Main Results

A basic tool for the forthcoming proofs are the following lemmata. The first is an application of the corresponding result in [12] and we will derive lemma 1 by using techniques developed in [5]. The second lemma states a purely point set topological fact and is attributed to ideas of Eells-Gleason-Palais (cf. [8]).

Lemma 1. Let X be a metric ANR, $F: X \to X$ a compact map and p a prime. Then

$$\Lambda(f,X) \equiv \Lambda(f^p,X) \bmod p \text{ .}$$

Proof. From [1] (cf. [5]) we find a U open in a linear normed space and mappings $r: U \to X$ and $s: X \to U$ such that $rs = id_X$. Granas in [5] defines $\Lambda(f,X) = \Lambda(sfr,U)$ and proves that sfr is a Lefschetz map. Observe that $(sfr)^p = sf^pr$, hence it suffices to show $\Lambda(F,U) \equiv \Lambda(F^p,U) \bmod p$, where $F = sfr: U \to U$ is a compact map. For $\varepsilon > 0$ sufficiently small there exists a finite polyhedron

$P_\varepsilon \subset U$ and an ε-approximation $F_\varepsilon: U \to U$ such that $F_\varepsilon(U) \subset P_\varepsilon$ and F_ε is homotopic with F. Let Φ denote either F_ε or F_ε^p and $\tilde{\Phi}$ the restriction of Φ to P_ε. Then the commutativity of the diagram

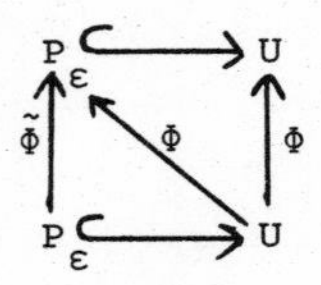

implies $\Lambda(\Phi,U) = \lambda(\tilde{\Phi},P_\varepsilon)$.

The proposition in [12] on the classical Lefschetz number gives

$$\lambda(\tilde{F}_\varepsilon,P_\varepsilon) \equiv \lambda(\tilde{F}_\varepsilon^p,P_\varepsilon) \bmod p .$$

Finally, $F \sim F_\varepsilon$ implies

$$\Lambda(F,U) = \Lambda(F_\varepsilon,U)$$

and

$$\Lambda(F^p,U) = \Lambda(F_\varepsilon^p,U) .$$

Applications of the 'mod p' theorem for the Lefschetznumber to transformations of prime period will be given in [14].

Lemma 2. Let X be a normal space, $A \subset X$ closed, $W \subset X$ open and $f: X \to X$ a mapping such that

(i) $A \subset W$

(ii) there is $m \in \mathbb{N}$ such that $f^m(W) \subset A$

(iii) $f(A) \subset A$

Then there is Y open in X such that $A \subset Y \subset W$ and $\mathrm{cl} f(Y) \subset Y$.

Proof. Choose open sets V_j, $j=0,\ldots,m-1$, such that

$$A \subset V_{m-1} \subset \mathrm{cl}V_{m-1} \subset V_{m-2} \subset \ldots \subset V_1 \subset \mathrm{cl}V_1 \subset V_0 = W$$

($\mathrm{cl}U$ denotes the closure of U) and set

$$Y = \bigcap_{j=0}^{m-1} f^{-j}(V_j) .$$

Equipped with these lemmata we are ready to give complete proofs for our main results.

Theorem 1. Let X be a metric ANR, $x_0 \in X$, $f: X \to X$ a compact map such that

(i) x_0 is a repulsive fixed point of f relative $U \in U(x_0)$ and $f(x) \neq x$ for all $x \in \partial U$,

(ii) there is $V \in U(x_0)$, $\mathrm{cl}V \subset U$, and $i: X\backslash V \to X$ induces isomorphisms in H_* .

Then $i(X,f,U) = 0$ and $i(X,f,X\backslash clU) = \Lambda(f,X)$.

If X is contractible we know $\Lambda(f,X) = 1$, thus the theorem provides the existence of a fixed point for f in $X\backslash clU$. In this sense the determination of repulsive fixed points might be useful to find further fixed points. Assumption (ii) in some sense is a geometric boundary condition, which for example is always satisfied if

- X is a compact manifold with boundary ∂X and $x_0 \in \partial X$, or
- X is a compact, convex, infinite dimensional subset of a Banach space and x_0 is arbitrary in X (cf. proof of corollary 1).

The following example shows that assumption (ii) cannot be dropped.

Example. Let D be the closed unit ball in the plane, $U = \{x\in D: ||x|| < 1/2\}$ and $f: D \to D$ be given by $f = gh$, where $h(x) = 2x$, if $||x|| \leq 1/2$, and $h(x) = x/||x||$, if $||x|| \geq 1/2$, and g is a non-periodic rotation of D . Then zero is a repulsive fixed point of f relative U and in view of the existence of non-repulsive fixed points also the only fixed point of f . Moreover, $i(D,f,U) = i(D,f,D) = \lambda(f,D) = 1$, from the additivity and normalization property of the index. Clearly $H_*(D\backslash U) = H_*(S^1) \neq H_*(D) = H_*$ (point).

Proof. (of theorem 1). We prove $\Lambda(f,X) = i(X,f,X\backslash clU)$, then $i(X,f,U) = 0$ is a consequence of the additivity property of the index. Assumption (ii) implies the commutativity of the diagram

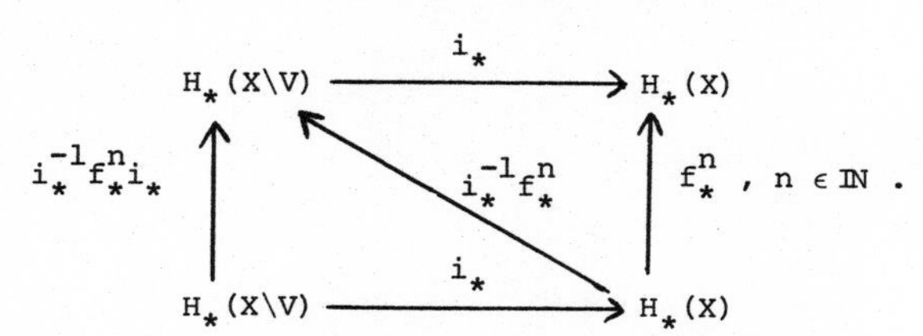

Since f_*^n is a Leray endomorphism we have that $i_*^{-1}f_*^n i_*$ is a Leray endomorphism and

$$\Lambda(f^n,X) = \Lambda(f_*^n) = \Lambda(i_*^{-1}f_*^n i_*) \ , \ n \in \mathbb{N} .$$

Since x_0 is repulsive we find $n_0 \in \mathbb{N}$ such that for all $n \geq n_0$

$$f^n(X\backslash V) \subset X\backslash U \subset X\backslash V .$$

Hence the diagram

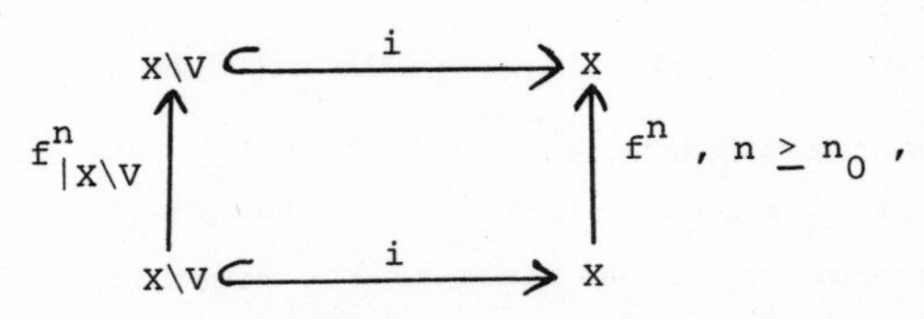

is commutative and $(f^n_{|X\backslash V})_* = i_*^{-1} f^n_* i_*$. Thus $\Lambda(f,X) = \Lambda((f^n_{|X\backslash V})_*)$, whenever $n \geq n_0$. We construct an open subset Y of X (i.e. Y is a metric ANR) such that

1) $f(Y) \subset Y$ and $f\colon Y \to Y$ is a compact map

2) $X\backslash V \subset Y$

3) there is $m_0 \in \mathbb{N}$ such that $f^n(Y) \subset X\backslash V$, whenever $n \geq m_0$.

Assume we have found Y with properties 1)-3). Then the commutativity of the diagram

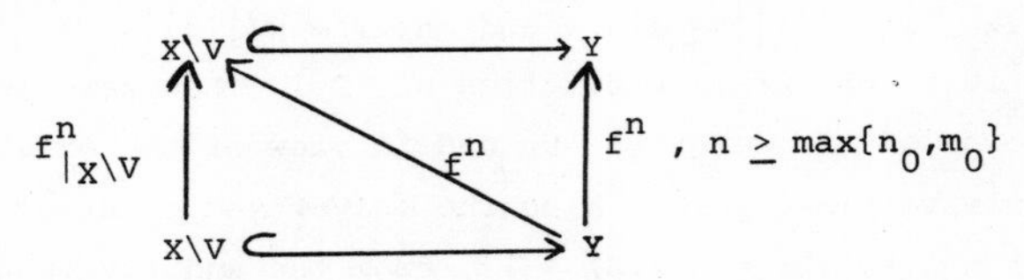

implies $\Lambda(f^n,Y) = \Lambda((f^n_{|X\backslash V})_*)$, whenever $n \geq \max\{n_0,m_0\}$. Hence $\Lambda(f^n,X) = \Lambda(f^n,Y)$. Lemma 1 implies $\Lambda(f,X) \equiv \Lambda(f^p,X) = \Lambda(f^p,Y) \equiv \Lambda(f,Y) \bmod p$, whenever p is a prime and p is sufficiently large. But this is only true if $\Lambda(f,X) = \Lambda(f,Y)$, which finishes the crucial step in the proof. Additivity, commutativity and the normalization property of the index finally imply

$$\Lambda(f,X) = \Lambda(f,Y) = i(Y,f,Y) = i(Y,f,X\backslash \mathrm{cl}V) = i(X,f,X\backslash \mathrm{cl}V) = i(X,f,X\backslash \mathrm{cl}U).$$

To construct Y set

$$A = X\backslash V \cup \mathrm{cl}f(X\backslash V) \cup \mathrm{cl}f^2(X\backslash V) \cup \ldots \cup \mathrm{cl}f^{n_0-1}(X\backslash V)$$

and observe that A is closed and $f(X) \subset A$. Since x_0 is a fixed point of f we have $x_0 \notin A$ and we find $V_1 \in U(x_0)$ such that $V_1 \subset V \subset U$ and $\mathrm{cl}V_1 \cap A = \emptyset$, i.e. $A \subset X\backslash \mathrm{cl}V_1$. Since x_0 is repulsive there is $m_0 \in \mathbb{N}$ such that

$$f^n(X\backslash V_1) \subset X\backslash U \subset X\backslash V \subset A \subset X\backslash \mathrm{cl}V_1 ,$$

whenever $n \geq m_0$. Setting $W = X\backslash \mathrm{cl}V_1$ we meet the assumptions of lemma 2 and find Y open in X such that $A \subset Y \subset W = X \backslash \mathrm{cl} V_1$ and $\mathrm{cl} f(Y) \subset Y$, i.e. $f\colon Y \to Y$ is a compact map.

As an immediate consequence of theorem 1 we obtain Browder-s result.

<u>Corollary 1</u>. Let X be a compact, convex, infinite dimensional subset of a Banach space and $f\colon X \to X$ a map. Then f has a non-repulsive fixed point.

<u>Proof</u>. Compactness of X implies that f has at most finitely many repulsive fixed points $\{x_1,\ldots,x_r\} = R$, since repulsive fixed points are isolated fixed points. If $R = \emptyset$ the conclusion follows from the Schauder fixed point theorem. From the Krein-Milman theorem we find extreme points $y_1,\ldots,y_r$ in X and from a theorem of Klee [6] we find a homeomorphism $H\colon X \to X$ such that $H(x_j) = y_j$. Now the y_j admit arbitrarily small open neighborhoods U_j such that $X \backslash U_j$ is contractible (cf. [2], lemma 2), thus assumption (ii) is satisfied for U_j, $j=1,\ldots,r$. Moreover, y_j is a repulsive fixed point of $F = HfH^{-1}$ relative U_j, if U_j is sufficiently small. Set $A = \mathrm{cl} \bigcup_{j=1}^{r} U_j$. Theorem 1 implies

$$i = \Lambda(F,X) = \Sigma_{j=1}^{r}\, i(X,F,U_j) + i(X,F,X\backslash A) = i(X,F,X\backslash A)$$

, i.e. F has a fixed point in $X\backslash A$, i.e. f has a fixed point in $H^{-1}(X\backslash A)$.

Notice that for example assumption (ii) is always satisfied if

- P is a normal cone in a Banach space,
- $f\colon P \to P$ is a compact map,
- 0 is a repulsive fixed point of f relative U and $f(x) \neq x$ for all $x \in \partial U$.

Since $\Lambda(f,P) = i(P,f,P\backslash \mathrm{cl} U) = 1$ we get the existence of a nontrivial fixed point for f. Generalizations of this observation will be given in [13].

For certain applications (cf. [9], [10]) it is of some importance to have an index characterization even in the case if f is not continuously defined in the repulsive fixed point. Such a characterization is provided by the next theorem and again x_0 being repulsive forces the map to have 'further' fixed points.

Theorem 2. Let X be a metric ANR, $x_0 \in X$, $f: X\setminus\{x_0\} \to X$ a compact map such that

(i) x_0 is a repulsive point for f relative $U \in U(x_0)$ and $f(x) \neq x$ for all $x \in \partial U$

(ii) for all $V \in U(x_0)$ $x_0 \notin \mathrm{cl} f(X\setminus V)$

(iii) there is $V_0 \in U(x_0)$ and $K \subset X$ such that $\mathrm{cl}V_0 \subset U$, $\tilde{H}_*(K) = 0$ and $X\setminus U \subset K \subset X\setminus V_0$.

Then $i(X,f,X\setminus \mathrm{cl}U) = 1$.

Proof. Since x_0 is repulsive there is $n_0 \in \mathbb{N}$ such that for all $n \geq n_0$

$$f^n(X\setminus \mathrm{cl}V_0) \subset f^n(X\setminus V_0) \subset X\setminus U \subset K \subset X\setminus V_0 .$$

Hence we have commutative diagrams

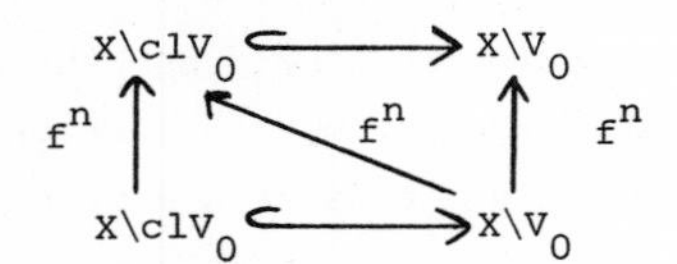
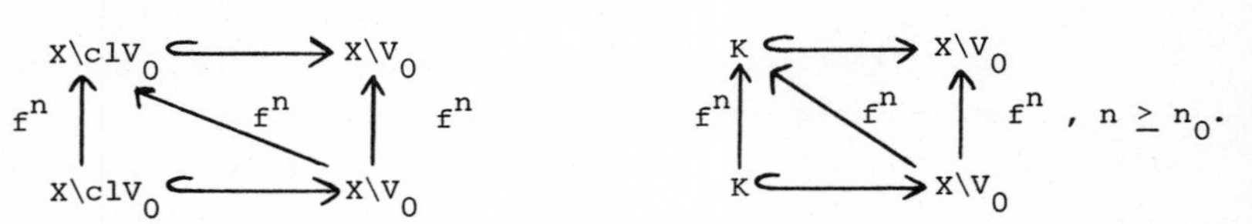

$X\setminus \mathrm{cl}V_0$ is open in X (i.e. $X\setminus \mathrm{cl}V_0$ is a metric ANR) and $\mathrm{cl}f^n(X\setminus \mathrm{cl}V_0) \subset \mathrm{cl}X\setminus U \subset X\setminus \mathrm{cl}V_0$ (i.e. f^n is a compact map on $X\setminus \mathrm{cl}V_0$). Thus $\Lambda(f^n, X\setminus \mathrm{cl}V_0)$ is defined and from the first diagram we obtain $\Lambda(f^n, X\setminus \mathrm{cl}V_0) = \Lambda(f^n, X\setminus V_0)$, whenever $n \geq n_0$. Commutativity of the second diagram implies $\Lambda(f^n,K) = \Lambda(f^n, X\setminus V_0)$, whenever $n \geq n_0$. But assumption (iii) implies $\Lambda(f^n,K) = 1$. Again we construct Y open in X with the properties 1) - 3) (cf. proof of theorem 1). To find $V_1 \in U(x_0)$ such that $\mathrm{cl}V_1 \cap A = \emptyset$ it suffices to know that $x_0 \notin A$, which follows in the present situation from assumption (ii) and an easy induction argument. Now we proceed according to the proof of theorem 1 and get $\Lambda(f^n,Y) = 1$, whenever $n \geq \max\{n_0,m_0\}$. Hence lemma 1 applied once again gives $\Lambda(f,Y) = 1$ and finally $i(X,f,X\setminus \mathrm{cl}U) = 1$.

Corollary 2. (cf. [9], theorem 1.1). Let X be a closed, convex infinite dimensional subset of a Banach space, $x_0 \in X$, $f: X\setminus\{x_0\} \to X$ a compact map and x_0 a repulsive point for f relative $U \in U(x_0)$ and $f(x) \neq x$ for all $x \in \partial U$. Then f has a fixed point in $X\setminus U$.

Proof. Choose a compact, convex, infinite dimensional set K

such that $x_0 \in K \subset X$. Then $P = \overline{\text{cov}}(K \cup f(X \setminus \{x_0\}))$ is compact, convex and infinite dimensional. Clearly x_0 is repulsive relative $U \cap P$, $f: P \setminus \{x_0\} \to P$ and assumption (ii) of theorem 2 is satisfied by compactness of P. By the same reasoning as in the proof of corollary 1 we may assume that x_0 is an extreme point of P, thus assumption (iii) of theorem 2 is satisfied. Hence theorem 2 implies $i(P,f,P \setminus \text{cl}(P \cap U)) = 1$.

Notice, that the assumption 'X is of infinite dimension' in corollary 2 is to guarantee that assumption (iii) in theorem 2 is satisfied. If X is of finite dimension and x_0 is an extreme point of X the conclusion of corollary 2 is true, because x_0 admits arbitrarily small neighborhoods with contractible complements.

<u>Theorem 3</u>. Let X be a metric ANR, $f: X \to X$ a compact map, $x_0 \in X$ such that

(i) x_0 is an attractive fixed point of f relative $U \in \mathcal{U}(x_0)$ and $f(x) \neq x$ for all $x \in \partial U$,

(ii) there is $W \in \mathcal{U}(x_0)$ and $K \subset X$ such that $W \subset K \subset U$, $\text{cl}W \subset U$ and $\tilde{H}_*(K) = 0$.

Then $i(X,f,U) = 1$.

<u>Proof</u>. Choose $V \in \mathcal{U}(x_0)$ such that $\text{cl}V \subset W$. Since x_0 is attractive there is $n_0 \in \mathbb{N}$ such that for all $n \geq n_0$ $f^n(U) \subset V$ and especially $\text{cl}f^n(W) \subset \text{cl}V \subset W$, i.e. f^n is a compact map on the metric ANR W. By assumption (ii) we have $f^n(K) \subset W$, whenever $n \geq n_0$, which again implies commutativity of the diagram

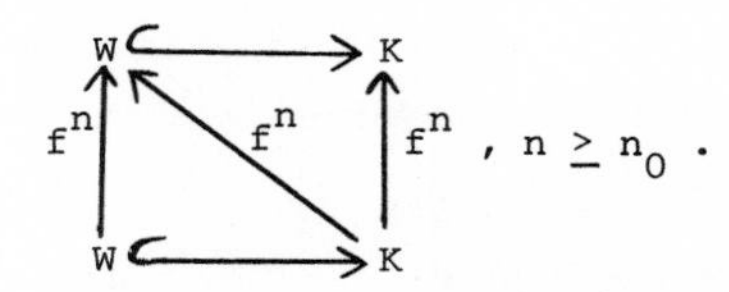

$, n \geq n_0$.

Hence $\Lambda(f^n,K)$ is defined and $\Lambda(f^n,W) = \Lambda(f^n,K)$. But $\Lambda(f^n,K) = 1$ for all $n \geq n_0$. Similar to the proofs of theorems 1 and 2 we construct an open set $Y \subset X$ such that

1) $Y \subset W$

2) $f(Y) \subset Y$ and $f: Y \to Y$ is a compact map

3) there is $m_0 \in \mathbb{N}$ such that for all $n \geq m_0$ $f^n(W) \subset Y$.

Assume we have found Y such that 1) - 3) is true. Then we proceed as in the proofs of theorems 1 and 2 to obtain $i(X,f,U) = 1$. The construction of Y in the present situation proceeds according to that in the previous proofs, but is a little more complicated. We set

$$V_1 = \bigcap_{j=0}^{n_0} f^{-j}(V)$$

and observe that $x_0 \in V_1$ and $f^n(V_1) \subset V$ for all $n \geq 0$. Since V_1 is an open neighborhood of x_0 we find $m_0 \in \mathbb{N}$ such that for all $n \geq m_0$

$$f^n(W) \subset f^n(U) \subset V_1 .$$

Now we set

$$A = \mathrm{cl}V_1 \cup \mathrm{cl}f(V_1) \cup \mathrm{cl}f^2(V_1) \cup \ldots \cup \mathrm{cl}f^{m_0-1}(V_1)$$

and observe that A is closed, $f(A) \subset A$. Since $f^n(V_1) \subset V$, $n \geq 0$, and $\mathrm{cl}V \subset W$ we see that $A \subset W$. Moreover $f^n(W) \subset V_1 \subset A$ for all $n \geq m_0$. Hence lemma 2 applies and we find Y open in X such that $A \subset Y \subset W$ and $\mathrm{cl}f(Y) \subset Y$.

Remark. In case X is a compact metric ANR assumption (ii) can be dropped. In this special case the general idea of a proof is the same as above except for the following:

- one constructs (cf. [4] lemma 2, and [8]) a compact metric ANR Y with the properties 1) - 3) (cf. proof of theorem 3)
- one observes that $C_\infty = \bigcap_{n>0} f^n(Y) = \{x_0\}$ is homologically trivial and applies a theorem of Deleanu [3], p. 224, to obtain $\lambda(f,Y) = 1$.

Finally we establish as an application of theorems 1 and 3 a bifurcation result in the spirit of Nussbaum-s ideas in [10].

Theorem 4. Let X be a metric ANR, J an open interval in $\mathbb{R}$, $F: X \times J \to X$ a compact map, $\tau: J \to X$ a continuous curve, $\tau(t) = x_t$, $t_0 \in J$, $x_{t_0} = x_0$ such that

(i) $F(x_t,t) = x_t$ for all $t \in J$,

(ii) x_t is a repulsive fixed point of F_t relative $U_t \in U(x_t)$, whenever $t < t_0$, and $F_t(x) \neq x$ for all $x \in \partial U_t$ and there is $V_t \in U(x_t)$, $\mathrm{cl}V_t \subset U_t$, such that $i: X \setminus V_t \to X$ induces isomorphisms in H_*,

(iii) x_t is an attractive fixed point of F_t relative $U_t \in U(x_t)$, whenever $t > t_0$, and $F_t(x) \neq x$ for all $x \in \partial U_t$ and there is $V_t \in U(x_t)$, $clV_t \subset U_t$, and $K_t \subset X$ such that $V_t \subset K_t \subset U_t$ and $\tilde{H}_*(K_t) = 0$.

Then (x_0,t_0) is a bifurcation point for $F(x,t) = x$.

Proof. Assume that (x_0,t_0) is not a bifurcation point, i.e. there is an open neighborhood Ω of (x_0,t_0) in $X\times J$, such that $F(x,t) \neq x$ for all $(x,t) \in \Omega$ such that $x \neq x_t$. Choose $W \in U(x_0)$ and a bounded interval $I \in U(t_0)$ such that

$$W\times ClI \subset \Omega \quad \text{and} \quad \tau(clI) \subset W .$$

This is possible, because τ is continuous and W is an open neighborhood of x_0 in X. Moreover observe that $\{x\in W: F_t(x)=x\}=x_t$, $t \in clI$, is compact. Hence the homotopy invariance of the index implies

$$i(X,F_{t_1},W) = i(X,F_{t_2},W) \text{ , where } t_1,t_2 \in clI \text{ and } t_1 < t_0 < t_2 .$$

Additivity of the index and theorems 1 and 3 imply

$$i(X,F_{t_1},W) = i(X,F_{t_1},U_{t_1}) = 0 \quad \text{and} \quad i(X,F_{t_2},W) = i(X,F_{t_2},U_{t_2}) = 1,$$

contradiction.

Acknowledgement

I would like to thank the Université de Montréal and Professor Andrzej Granas for hospitality during a visit in which I prepared this manuscript.

References

[1] Arens, R.F., Eells, J., On embedding uniform and topological spaces. Pacific J. Math. 6, (1956) 397-403.

[2] Browder, F.E., Another generalization of the Schauder fixed point theorem, Duke Math. J. 32 (1965), 399-406.

[3] Deleanu, A., Une généralisation du théorème du point fixe de Schauder, Bull. Soc. Math. France 89 (1961), 223-226.

[4] Fenske, C.C., Peitgen, H.O., Repulsive fixed points of multivalued transformations and the fixed point index, Math. Ann.

[5] Granas, A., The Leray-Schauder index and the fixed point theory for arbitrary ANR-s, Bull. Soc. Math. France 100 (1972), 209-228.

[6] Klee, V., Some topological properties of convex sets. Trans. Amer. Math. Soc. 78 (1955), 30-45.

[7] Leray, J., Théorie des points fixes: indice total et nombre de Lefschetz, Bull. Soc. Math. France 87 (1959), 221-233.

[8] Nussbaum, R.D., Asymptotic fixed point theorems for local condensing maps, Math. Ann. 191 (1971), 181-195 .

[9] Nussbaum, R.D., Periodic solutions of some nonlinear autonomous functional differential equations, Ann. Mat. Pura Appl. 101 (1974) 263-306.

[10] Nussbaum, R.D., A global bifurcation theorem with applications to functional differential equations, J. Functional Analysis 19 (1975), 319-338.

[11] Peitgen, H.O., Asymptotic fixed point theorems and stability, J. Math. Analysis Appl. 47 (1974), 32-42.

[12] Peitgen, H.O., On the Lefschetz number for iterates of continuous mappings, Proc. Amer. Math. Soc.

[13] Peitgen, H.O., Attractivity and ejectivity in cones of Banach spaces.

[14] Peitgen, H.O., Transformations of prime period and the Lefschetz number.

[15] Steinlein, H., Über die verallgemeinerten Fixpunktindizes von Iterierten verdichtender Abbildungen, Manuscripta math. 8 (1972) 251-266.

[16] Zabreiko, P.P., Krasnosel'skii, M.A., Iterations of operators and the fixed point index. Doklady Akad. Nauk SSSR 196, (1971), 1006-1009 = Soviet Math. Dokl. 12 (1971), 294-298.

Institut für Angewandte Mathematik
der Universität Bonn
D-53 Bonn
Wegelerstrasse 6
Germany F.R.

ON THE RELATIONSHIP OF A-PROPERNESS TO MAPPINGS OF MONOTONE TYPE WITH APPLICATIONS TO ELLIPTIC EQUATIONS*

W. V. Petryshyn[+]

Introduction

Unless stated otherwise X and Y will always denote separable real normed spaces with X^* and Y^* denoting their respective dual spaces, $T: X \to Y$ a nonlinear mapping, $\Gamma = \{E_n, V_n; F_n, W_n\}$ an admissible approximation scheme for the equation

$$\dot{T}(x) = f \quad , \quad (x \in X,\ f \in Y) \ , \tag{1}$$

and

$$T_n(u_n) = W_n(f) \quad , \quad (u_n \in E_n \ , \ W_n f \in F_n) \tag{2}$$

a sequence of approximate equations associated with Equation (1) by means of the scheme Γ where $T_n \equiv W_n T V_n : E_n \to F_n$ is a finite dimensional approximation to T. In a number of earlier papers (see [27] for references) the author studied the approximation-solvability of Equation (1) under the conditions that T is A-proper and either a-stable or K-coercive[(1)] (see [23, 27]) or T is pseudo-A-proper and K-coercive or norm-coercive (see [25, 27]).

The purpose of this paper is two-fold. In Section 1 we first study (see Theorems 1, 2 and 3 below) the approximation-solvability of Equation (1) under the conditions that T is <u>A-proper</u> with respect to (w.r.t.) a general (not necessarily projectionally complete) scheme Γ and T satisfies <u>condition</u> (+) given by:

(+) If $\{x_k\}$ is any sequence such that $T(x_k) \to g$ for some g in Y, then $\{x_k\}$ is bounded.

The obtained results (e.g. Theorem 2) are then used to establish the solvability of Equation (1) for each f in Y (see Theorem 4 below) for the case when T is a <u>uniform limit</u> of a special sequence of A-proper mappings w.r.t. Γ under the additional <u>condition</u> (++)

on T given by:

(++) If $\{x_k\}$ is any bounded sequence such that $T(x_k) \to g$ for some g in Y, then there is an $x \in X$ such that $T(x) = g$.

In case X is reflexive and $K: X \to Y^*$ satisfies suitable conditions, Theorem 4 is then used to deduce the solvability of Equation (1) for each f in Y (see Theorem 5 below) for the case when $T: X \to Y$ is K-quasimonotone or of type (P) in the sense of [10, 16] satisfies conditions (+) and (++). As a special case of Theorem 5 we deduce the basic results of Fitzpatrick [15] and Calvert-Webb [10] for quasimonotone mappings $T: X \to X^*$ as well as some earlier results of Minty [21], Browder [2, 4, 5], Brezis [1], Rockafellar [30] and others for monotone, pseudo-monotone and semimonotone mappings. The above authors used different arguments and established their results for spaces which need not be separable.

Let us add that in case X and Y are Banach spaces with Schauder bases, Theorems 1 and 4 were stated by the writer in his paper [27] without detailed proofs.

In section 2 we apply the results of Section 1 on A-proper mappings to the approximation-solvability of quasilinear elliptic boundary value problems of order 2 to obtain strong solutions in the Sobolev space W_2^2. This continues our study of the solvability of differential equations initiated in [27] by means of the A-proper mapping theory.

Section 1

Let $\{E_n\}$ and $\{F_n\}$ be two sequences of oriented finite dimensional spaces and let $\{V_n\}$ and $\{W_n\}$ be two sequences of continuous linear mappings with V_n mapping E_n into X and W_n mapping Y onto F_n.

Remark 1. For the sake of notational simplicity we use the same symbol $\|\ \|$ to denote the norms in the respective spaces X, Y, E_n and F_n. We hope that at each step it will be clear to the reader which norm is meant. We also use the symbols "$\to$" and "$\rightharpoonup$" to denote strong and weak convergence, respectively.

Definition 1. A quadruple of sequences $\Gamma = \{E_n, V_n; F_n, W_n\}$ is said to be an admissible scheme for Equation (1) if $\dim E_n = \dim F_n$ for each n, V_n is injective, $\operatorname{dist}(x, V_n E_n) \to 0$ as $n \to \infty$ for each $x \in X$, and $\{W_n\}$ is uniformly bounded.

Note that Definition 1 does not require that E_n and F_n be subspaces of X and Y respectively nor that V_n and W_n be linear projections. The following examples will illustrate the generality of Definition 1. Others will be given later.

Let $\{X_n\}$ be a sequence of oriented finite dimensional subspaces of X such that $\operatorname{dist}(x,X_n) \to 0$ as $n \to \infty$ for each x in X and let V_n be a linear injection of X_n into X.

(a) Let $\{Y_n\}$ be a sequence of finite dimensional oriented subspaces of Y such that $\dim Y_n = \dim X_n$ and let Q_n be a continuous linear map of Y onto Y_n such that $||Q_n|| \leq M$ for all n and some $M > 0$. Then $\Gamma_a = \{X_n,V_n; Y_n,Q_n\}$ is admissible for $T: X \to Y$.

(b) If $Y = X$, $Y_n = X_n$ and $W_n = P_n$, where P_n is a projection of X onto X_n such that $P_n x \to x$ for each $x \in X$ and $||P_n|| \leq M_0$ for all n, then $\Gamma_b = \{X_n,V_n; X_n,P_n\}$ is an admissible projection scheme for $T: X \to X$. Note that when X is complete, then the assumption $||P_n|| \leq M_0$ is superfluous.

(c) If $Y = X^*$, $Y_n = R(P_n^*)$, $V_n = P_n|X_n = I_n$ and $W_n = P_n^*$, then $\Gamma_c = \{X_n,P_n; Y_n,P_n^*\}$ is an admissible projection scheme for $T: X \to X^*$.

(d) If $Y = X^*$, $Y_n = X_n^*$ and $W_n = V_n^*$, then $\Gamma = \{X_n,V_n; X_n^*,V_n^*\}$ is an admissible injection scheme for $T: X \to X^*$.

We add that the scheme Γ_d, which proved to be particularly useful (see [20, 4, 5]) for the approximation-solvability of boundary value problems for differential equations, always exist when X is separable. Example (c) shows that a projection scheme could be admissible for $T: X \to X^*$ without being projectionally complete for the pair (X,X^*) (i.e., such that $P_n x \to x$ for $x \in X$ and $P_n^* g \to g$ for g in X^*).

The class of operators T studied in this paper is given by the following definition first introduced by the author in [22] (see [27] for the historical development of the theory of A-proper mappings).

<u>Definition 2</u>. $T: X \to Y$ is said to be A-proper w.r.t. $\Gamma = \{E_n,V_n;F_n,W_n\}$ if $T_n \equiv W_n T V_n : E_n \to F_n$ is continuous for each n and if for any sequence $\{u_{n_j} | u_{n_j} \in E_{n_j}\}$ such that $\{V_{n_j} u_{n_j}\}$ is bounded in X and

$||T_{n_j}(u_{n_j}) - W_{n_j}(y)|| \to 0$ as $j \to \infty$ for some y in Y, there exists a subsequence $\{u_{n_{j(k)}}\}$ and $x_0 \in X$ such that $V_{n_{j(k)}} u_{n_{j(k)}} \to x_0$ and $Tx_0 = y$.

The following examples (for other and more complicated examples and references see [27]) illustrate the generality of the class of A-proper mappings. First let us note that if T is A-proper w.r.t. Γ and $C: X \to Y$ is compact (i.e., T is continuous and maps bounded sets in X into relatively compact sets in Y), then $T+C$ is A-proper w.r.t. Γ.

Example 1. If $C: X \to X$ is compact, then $T = I-C$ is A-proper w.r.t. Γ_b.

Example 2. If X is complete, $S: X \to X$ strictly contractive, then $T = I-S-C$ is A-proper w.r.t. Γ_b provided $||P_n|| = 1$ for all n.

Example 3. If X is reflexive and $T: X \to X^*$ is strongly monotone, i.e., $(Tx-Ty, x-y) \geq c||x-y||^2$ for $x,y \in X$ and some $c > 0$, and either continuous, demicontinuous, or weakly continuous, then T is A-proper w.r.t. Γ_c and Γ_d provided that T is bounded. The last condition could be omitted if Γ_c is projectionally complete for (X,X^*).

Example 4. If X is reflexive and $T: X \to X^*$ is bounded, demicontinuous and satisfies condition (s) of Browder [5] (i.e., T is such that if $x_k \rightharpoonup x$ in X and $(Tx_k-Tx, x_k-x) \to 0$, then $x_k \to x$ in X), then T is A-proper w.r.t. Γ_d.

We add in passing that if X and $T: X \to X^*$ satisfy the conditions of Example 3 or of Example 4 and $C: X \to X^*$ is weakly continuous and orthogonal (i.e. $(Cx,x) = 0$ for all $x \in X$), then $T + \mu C$ is A-proper w.r.t. Γ_c and Γ_d for each $\mu \geq 0$. This observation is important since the time independent Navier-Stokes equation can be studied in the framework of A-proper mapping theory. We also add that the class of P_γ-compact mappings and, in particular, translations of ball-condensing mappings, form a subclass of A-proper mappings.

To state our results precisely we need the following.

Definition 3. For a given f in Y, Equation (1) is said to be strongly (resp. feebly) approximation-solvable w.r.t. Γ if there exists an integer $N_f \geq 1$ such that Equation (2) has a solution $u_n \in E_n$ for each $n \geq N_f$ with the property that $V_n u_n \to x_0$ in X (resp., $V_{n_j} u_{n_j} \to x_0$ for some subsequence $\{u_{n_j}\}$ of $\{u_n\}$) and $T(x_0) = f$.

Note that the approximation-solvability of Equation (1) in strong or feeble sense implies the solvability of Equation (1) but the converse is not true in general. Our first result is the following theorem which extends Theorem 3.1M and Corollary 3.2E stated in [27] essentially for the case of Banach spaces X and Y with Schauder bases.

Theorem 1. Let (X,Y) be normed spaces with an admissible scheme Γ given by Definition 1. Let K be a map of X into Y^* and K_n a map of E_n into F_n^* such that $Kx = 0$ implies $x = 0$ and for each n

(C1) $(W_n(g), K_n(u)) = (g, KV_n(u))$ for $u \in E_n$ and $g \in Y$. Let M_n be a linear isomorphism of E_n onto F_n such that

(C2) $(M_n(u), K_n(u)) > 0$ for $u \in E_n$ with $u \neq 0$.

Let $T: X \to Y$ be A-proper w.r.t. Γ and satisfy one of the following two conditions

(i) there is an $r_0 > 0$ such that T is odd on $X - B(0,r_0)$, where $B(0,r_0) = \{x \in X \mid \|x\| < r_0\}$.

(ii) there is an $r_0 > 0$ such that $(Tx, Kx) \geq 0$ for $\|x\| > r_0$.

Then, if T satisfies condition (+), Equation (1) is feebly approximation-solvable w.r.t. Γ for each $f \in Y$. Equation (1) is strongly approximation-solvable if it is uniquely solvable for a given f.

We shall deduce Theorem 1 as a corollary of the following slightly more general theorem which will prove to be useful in our study of the solvability of Equation (1) in case T is a uniform limit of a suitable sequence of A-proper mappings.

Theorem 2. Suppose that all the conditions of Theorem 1 hold except that condition (+) is replaced by the hypothesis:

(H1): To each f in Y there correspond an $r > r_0$ and $\alpha > 0$ (depending on f) such that

$$\|Tx - tf\| \geq \alpha \quad \text{for} \quad x \in \partial B(0,r) \quad \text{and} \quad t \in [0,1] . \tag{3}$$

Then the conclusions of Theorem 1 hold.

<u>Proof</u>: The A-properness of T and (H1) imply the existence of an integer n_0 and a number $\gamma > 0$ (depending on f) such that

$$||T_n(u)-tW_nf|| \geq \gamma \quad \text{for} \quad n \geq n_0 \,,\ t \in [0,1] \ \text{and} \ u \in \partial B_n \,, \tag{4}$$

where $B_n \equiv V_n^{-1}(B(0,r)) \equiv \{u \in E_n | V_n(u) \in B(0,r)\}$ and $V_n^{-1}(clB(0,r))$ are open and closed sets in E_n, respectively, with $B_n \cap \partial B_n = \emptyset$, $cl(B_n) \subseteq V_n^{-1}(cl(B(0,r))$ and the boundary $\partial B_n \subset V^{-1}(\partial B(0,r))$ for each n. Moreover, for each fixed n, B_n is bounded, convex, and symmetric about $0 \in E_n$. Now suppose that our assertion is false. Then there exists a sequence $\{n_j\}$ of positive integers with $n_j \to 0$ and sequences $\{t_{n_j}\} \subset [0,1]$ and $u_{n_j} \in \partial B_{n_j}$ with $||V_{n_j}(u_{n_j})|| = r$ and $t_{n_j} \to t_0 \in [0,1]$ such that $||T_{n_j}(u_{n_j})-t_{n_j}W_{n_j}f|| \to 0$ as $j \to \infty$. Hence $T_{n_j}(u_{n_j})-W_{n_j}(t_0f) = T_{n_j}(u_{n_j})-t_{n_j}W_{n_j}f+(t_{n_j}-t_0)W_{n_j}(f) \to 0$ as $j \to \infty$ since $\{W_{n_j}(f)\}$ is bounded. Consequently, by the A-properness of T w.r.t. Γ, there exists a subsequence $\{u_{n_{j(k)}}\}$ and x_0 in X such that $V_{n_{j(k)}}(u_{n_{j(k)}}) \to x_0$ in X and $T(x_0) = t_0f$ with $||x_0|| = r$, in contradiction to (3).

<u>Case (i)</u>. If T is odd on $X-B(0,r)$, then so is the mapping T_n on ∂B_n and thus it follows from (4) and the homotopy and Borsuk theorems for the Brouwer degree that for each fixed $n \geq n_0$ we have

$$0 \neq \deg(T_n,B_n,0) = \deg(T_n-W_nf,B_n,0) \,.$$

Hence, for each such n, there exists $u_n \in B_n$ such that $T_n(u_n) = W_nf$. Since T is A-proper, there exists a subsequence $\{u_{n_j}\}$ and an $x_0 \in B(0,r)$ such that $V_{n_j}(u_{n_j}) \to x_0$ in X and $T(x_0) = f$, i.e., Equation (1) is feebly approximation-solvable for each $f \in Y$.

<u>Case (ii)</u>. If $(Tx,Kx) \geq 0$ for $||x|| \geq r_0$, then we consider the homotopy $H_n(t,x)\colon [0,1] \times cl(B_n) \to F_n$ defined for each fixed $n \geq n_0$ by

$$H_n(t,x) = tT_n(u) + (1-t)M_n(u) \,, \quad (u \in Cl(B_n) \,,\ t \in [0,1]) \,. \tag{5}$$

We claim that $H_n(t,x) \neq 0$ for $x \in \partial B_n$ and $t \in [0,1]$. Indeed, if this were not the case, then there would exist $u_0 \in \partial B_n$ and $t_0 \in [0,1]$ such that

$$H_n(t_0,x_0) = t_0 T_n(u_0) + (1-t_0)M_n(u_0) = 0 .$$

It follows from (4) that $t_0 \neq 1$ and from (C2) that $t_0 \neq 0$. Thus $t_0 \in (0,1)$, $V_n(u_0) \in \partial B(0,r)$ and from (C1) and (C2) we find that

$$(TV_n(u_0),KV_n(u_0)) = (T_n(u_0),K_n(u_0)) = -((1-t_0)/(t_0)(M_n(u_0),K_n(u_0)) < 0,$$

in contradiction to our hypothesis (ii). Thus $H_n(t,x) \neq 0$ and so, by the homotopy theorem for the Brouwer degree, we find that

$$\deg(T_n,B_n,0) = \deg(M_n,B_n,0) \neq 0$$

for each $n \geq n_0$ since M_n is a linear isomorphism of E_n onto F_n. It follows from this and (4) that $\deg(T_n - W_n f, B_n, 0) \neq 0$ for each $n \geq n_0$ and so there exists $u_n \in B_n$ such that $T_n(u_n) = W_n f$. This and the A-properness of T w.r.t. Γ imply the existence of a subsequence $\{u_{n_j}\}$ and $x_0 \in cl(B(0,r))$ such that $V_{n_j}(u_{n_j}) \to x_0$ and $T(x_0) = f$. This proves the first and the main part of Theorem 2.

To prove the last assertion of Theorem 2 we note that, by what has been proved above, for each f in Y there exists a sequence $\{u_n | u_n \in E_n\}$ of solutions of Equation (2) and a strong limit point x_0 of $\{V_n(u_n)\}$ in X such that $T(x_0) = f$. Suppose that for a given f in Y, Equation (1) has at most one solution. Then x_0 is the unique solution of Equation (1) and, therefore, $V_n(u_n) \to x_0$ in X. Indeed, if not, then there would exist a subsequence $\{u_{n_k}\}$ such that $||V_{n_k}(u_{n_k}) - x_0|| \geq \varepsilon$ for all k and some $\varepsilon > 0$. But $T_{n_k}(u_{n_k}) = W_{n_k}(f)$ for each k and therefore, by the A-properness of T w.r.t. Γ, there exists a subsequence $\{u_{n_{k(i)}}\}$ and $x_0' \in X$ such $V_{n_{k(i)}}(u_{n_{k(i)}}) \to x_0'$ as $i \to \infty$ and $T(x_0') = f$ with $x_0' \neq x_0$. This contradiction establishes the last assertion of Theorem 2. Q.E.D.

Proof of Theorem 1: To prove Theorem 1, it suffices to show that the assumed condition (+) implies the validity of the hypothesis (H1) of Theorem 2.

Now, if for some f in Y the hypothesis (H1) fails to hold, then we could find sequences $\{t_{n_j}\} \subset [0,1]$ and $\{x_{n_j}\} \subset X$ with

$t_{n_j} \to t_0 \in [0,1]$ and $||x_{n_j}|| \to \infty$ as $i \to \infty$ such that $T(x_{n_j}) - t_{n_j} f \to 0$. This implies that $T(x_{n_j}) \to t_0 f$ as $i \to \infty$ with $\{x_{n_j}\}$ unbounded, in contradiction to condition (+). Thus condition (+) implies (H1) and so Theorem 1 follows from Theorem 2. Q.E.D.

Remark 2. Note that condition (+) is equivalent to the requirement that $T^{-1}(Q) = \{x \in X \mid Tx \in Q\}$ be bounded whenever Q is relatively compact in Y. We add that condition (+) is implied by any one of the following conditions which have been used by a number of authors (see [4, 15, 29, 30]) in their study of Equation (1) involving operators of monotone type and ball-condensing type:

Condition (1+): $\frac{(Tx,Kx)}{||Kx||} \to \infty$ as $||x|| \to \infty$ (i.e., T is K-coercive).

Condition (2+): $||Tx|| \to \infty$ as $||x|| \to \infty$ (i.e., T is norm - coercive).

Condition (3+): $||Tx|| + \frac{(Tx,Kx)}{||K(x)||} \to \infty$ as $||x|| \to \infty$.

Condition (4+): $T(tx) = t^{\alpha} T(x)$ for $||x|| \geq r_0$ and $t > 1$ and $0 \notin T(\partial B(0,r_0))$.

Consequently, Theorem 1 remains valid if condition (+) is replaced by any one of the above conditions. Thus, a number of results obtained earlier by the writer and other authors can be deduced from Theorem 1.

Let us observe that Theorems 1 and 2 allow us to study the approximation-solvability of the perturbed equations of the form

$$T(x) \equiv A(x) + C(x) = f \quad , \quad (x \in X \; , \; f \in Y) \; , \tag{6}$$

where $C: X \to Y$ is relatively compact and A is A-proper w.r.t. Γ. In this case, conditions (+) and (ii) of Theorem 1 are implied, for example, by the

Assumption (a): A is K-coercive and $(Cx,Kx) \geq -c_0 ||Kx||$ for all $x \in X$ and some $c_0 > 0$.

Consequently, under Assumption (a), Equation (6) is feebly approximation-solvable w.r.t. Γ for each $f \in Y$.

The above result can be somewhat strengthened if we assume that A is odd.

Theorem 3. Suppose $A: X \to Y$ is A-proper w.r.t. Γ, odd on $X-B(0,r_0)$, and satisfies condition (+). Suppose $C: X \to Y$ is a bounded map such that

(C3) $T_t = A + tC$ is A-proper w.r.t. Γ for each $t \in [0,1]$.

(C4) To each $f \in Y$ there corresponds a number $c = c(||f||) > 0$ such that if the equation $Ax + tCx = f$ holds for some $x \in X$ and $t \in [0,1]$, then $||x|| \leq c(||f||)$.

Then Equation (6) is feebly approximation-solvable for each $f \in Y$ and strongly approximation-solvable if it is uniquely solvable.

Proof: Let f be an arbitrary but fixed element in Y. It follows from (C4) that there exists a number $r_1 > c(||f||)$ such that

$$T_t(x) = A(x) + tC(x) - f \neq 0 \quad \text{for} \quad ||x|| \geq r_1 \quad \text{and} \quad t \in [0,1]. \tag{7}$$

On the other hand, since A satisfies condition (+), there exists a number $r \geq \max\{r_0, r_1\}$ and $\alpha > 0$.

$$||A(x) - tf|| \geq \alpha \quad \text{for} \quad x \in \partial B(0,r) \quad \text{and} \quad t \in [0,1]. \tag{8}$$

Since A is odd on $X-B(0,r_0)$ and (8) holds for $r > r_0$, it follows from Theorem 1 in Browder-Petryshyn [9], concerning the generalized degree for A-proper maps, which is also valid for noncomplete spaces, that

$$0 \notin \mathrm{Deg}(A, B(0,r), 0) = \mathrm{Deg}(A-f, B(0,r), 0).$$

On the other hand, in view of the same Theorem 1 in [9], condition (C3) and (7) imply that

$$\mathrm{Deg}(A-f, B(0,r), 0) = \mathrm{Deg}(A+C-f, B(0,r), 0).$$

Hence, by the definition of the generalized degree (see [9]), there exists an integer $n_0 \geq 1$ such that $\deg(A_n + c_n - W_n f, B_n, 0) \neq 0$ for each $n \geq n_0$ and so for each such n there exists $u_n \in B_n$ such that $A_n(u_n) + C_n(u_n) - W_n f = 0$. Hence, by the A-properness of $A+C$, there exists a subsequence $\{u_{n_j}\}$ and $x_0 \in B(0,r)$ such that $V_{n_j}(u_{n_j}) \to x_0$ and $A(x_0) + C(x_0) = f$. The last assertion follows from the assumption of the uniqueness. Q.E.D.

Remark 3. Note that hypothesis (C3) always holds when C is compact. However, a remark following Example 4 shows that condition (C3) may hold for noncompact C.

Remark 4. It is useful to observe that condition (C4) of Theorem 3 is implied by the following:

Condition (C4'): $||Ax|| - ||Cx|| \to \infty$ as $||x|| \to \infty$.

The last condition holds, in particular, (see [34]) when $||Ax|| \to \infty$ as $||x|| \to \infty$ and $||Cx|| \leq \alpha_1 \; ||Ax|| + \alpha_2$ with $\alpha_1 \in (0,1)$ and $\alpha_2 > 0$.

If $||Ax|| \geq \beta||x||$ for $x \in X$ and $C: X \to Y$ is quasi-bounded with its quasinorm

$$|C| = \inf_{0\leq q<\infty} \{\sup_{||x||\geq q} ||Cx||/||x||\} < \beta , \tag{8}$$

then condition (C4') also holds. The latter always holds if C is asymptotically zero, i.e.,

$$||Cx||/||x|| \to 0 \quad \text{as} \quad ||x|| \to \infty \quad . \tag{9}$$

Using Theorem 2 it is now easy to establish a surjectivity theorem for Equation (1) when T is a uniform limit of a suitable sequence of A-proper mappings. Theorem 4 below extends Theorems 5.3F and 5.4B stated by the author in [27] (with only an indication of the proof) for Banach spaces with Schauder bases.

Theorem 4. Let (X,Y) , Γ, K, K_n and M_n satisfy the conditions of Theorem 1. Suppose $T: X \to Y$ satisfies condition (++) and $T_n : E_n \to F_n$ is continuous for each n . Suppose also that there exists a bounded map $G: X \to Y$ such that

$T_\mu = T + \mu G$ is A-proper w.r.t. Γ for each $\mu > 0$. (4a)

Suppose further that any one of the following two conditions holds:

(i) there is $r_0 > 0$ such that T and G are odd on $X-B(0,r_0)$.

(ii) there is $r_0 > 0$ such that $(Tx,Kx) \geq 0$ and $(Gx,Kx) \geq 0$ for $||x|| \geq r_0$.

Then, if T satisfies condition (+), $T(X) = Y$.

Proof: It was shown above that condition (+) implies the existence of $r \geq r_0$ and $\alpha > 0$ such that the inequality (3) holds for $x \in \partial B(0,r)$ and $t \in [0,1]$. In view of this and the boundedness of F , we can choose $\mu_0 > 0$ such that

$$||T_\mu(x)-tf|| \geq \frac{\alpha}{2} \quad \text{for} \quad x \in \partial B(0,r),\ t \in [0,1],\ \mu \in (0,\mu_0) \ . \tag{10}$$

Applying Theorem 2 to the equation $T_{\mu_k}(x) = T(x) + \mu_k C(x) = f$, where $\mu_k \in (0,\mu_0)$ is such that $\mu_k \to 0$ as $k \to \infty$, in either case we find an element $x_k \in B(0,r)$ such that $T(x_k) + \mu_k G(x_k) = f$ for each k . Since $\mu_k \to 0$ as $k \to \infty$ and $\{G(x_k)\}$ is bounded, it follows that $T(x_k) \to f$ as $k \to \infty$. This and condition (++) imply the existence of an $x \in X$ such that $T(x) = f$ for any given $f \in Y$. Q.E.D.

Remark 5. Theorem 4 remains valid if instead of condition (+) we assume that T satisfies any one of the conditions mentioned in Remark 2. Note that condition (++) certainly holds if $T(cl(B(0,r))$ is closed in Y for each $r > 0$.

In order to deduce from our Theorems 1 to 4 some general constructive and/or existence results for mappings $T: X \to Y$ of K-monotone type we need some further definitions and establish two lemmas.

Definition 4. In analogy with Browder [5] we say that $T: X \to Y$ satisfies condition (KS) if $x_k \rightharpoonup x$ in X and $(Tx_k - Tx, K(x_k - x)) \to 0$ imply that $x_k \to x$ in X . T satisfies condition (KS_+) if $x_k \rightharpoonup x$ in X and $\limsup (Tx_k - Tx, K(x_k - x)) \leq 0$ imply that $x_k \to x$ in X .

Our first lemma extends the result of Browder [5] to certain demicontinuous mappings $T: X \to Y$ satisfying condition (KS) .

Lemma 1. Suppose X is reflexive and the scheme $\Gamma_a = \{X_n, V_n; Y_n, Q_n\}$ is admissible for $T: X \to Y$. Let K be a bounded mapping of X into Y^* such that

(b1) $Kx = 0$ implies $x = 0$, K is positively homogeneous of order $\alpha > 0$ and $R(K)$ is dense in Y^* .

(b2) $(Q_n w, Kx) = (w, Kx)$ for all $x \in X_n$ and $w \in Y$.

(b3) K is weakly continuous at 0 and uniformly continuous on closed balls in X .

(b4) If $\{x_j\}$ is a bounded sequence in X such that $\{(Tx_j, Kx_j)\}$ is also bounded, then $\{Tx_j\}$ is bounded.

Then, if T is demicontinuous and satisfies condition (KS), T is A-proper w.r.t. Γ_a .

Proof: Let $\{x_{n_j} \mid x_{n_j} \in X_{n_j}\}$ be any bounded sequence such that $\|Q_{n_j} T_{n_j} - Q_{n_j} g\| \to 0$ for some g in Y . First note that, in view of (b2), it follows from the equality

$$(Tx_{n_j},Kx_{n_j}) = (Q_{n_j}Tx_{n_j}-Q_{n_j}g,Kx_{n_j}) + (g,Kx_{n_j})$$

that $\{(Tx_{n_j},Kx_{n_j})\}$ is bounded. Hence, by (b4), the sequence $\{Tx_{n_j}\}$ is bounded. Now, since X is reflexive and $\{x_{n_j}\}$ is bounded, we may assume that $x_{n_j} \rightharpoonup x_0$ in X. Since $\text{dist}(x_0,X_n) \to 0$, there is $y_n \in X_n$ such that $y_n \to x_0$ in X. Let $B(0,r)$ be a ball in X such that x_0, $\{x_{n_j}\}$ and $\{y_n\}$ lie in $cl(B(0,r))$. It follows from the fact that $x_{n_j}-y_{n_j} \rightharpoonup 0$, condition (b2), and the weak continuity at 0 of K that

$$(Tx_{n_j}-g,K(x_{n_j}-y_{n_j})) = (Q_{n_j}Tx_{n_j}-Q_{n_j}g,K(x_{n_j}-y_{n_j})) \to 0 .$$

The above relation implies that

$$(Tx_{n_j},K(x_{n_j}-y_{n_j})) \to (g,K(0)) = 0 \quad \text{as} \quad j \to \infty .$$

Now, since $K(tx) = t^{\alpha}K(x)$ for $x \in X$ and $t > 0$ and $\{x_{n_j}\}$, $\{y_{n_j}\}$ and x_0 lie in $cl(B(0,r))$ for each j, we have the relation

$$\begin{aligned}(Tx_{n_j},K(x_{n_j}-x_0)) &= (Tx_{n_j},K(x_{n_j}-y_{n_j})) + \\ &2\,(Tx_{n_j},K(\tfrac{1}{2}(x_{n_j}-x_0))-K(\tfrac{1}{2}(x_{n_j}-y_{n_j})))\end{aligned} \tag{11}$$

with elements $\frac{1}{2}(x_{n_j}-x_0)$ and $\frac{1}{2}(x_{n_j}-y_{n_j})$ lying in $cl(B(0,r))$. For each $t > 0$, define the function $\psi(t)$ as in [18] by

$$\psi(t) = \sup\{||Kx-Ky|| \;\big|\; ||x-y|| \le t \,,\; x,y \in cl(B(0,r))\} .$$

Since K is uniformly continuous on $cl(B(0,r))$, the function $\psi(t)$ is nondecreasing in t, $\psi(t) \to 0$ as $t \to 0$ and

$$||Kx-Ky|| \le \psi(||x-y||) \quad \text{for} \quad x,y \ \text{in} \ cl(B(0,r)) . \tag{12}$$

Since $\frac{1}{2}(x_{n_j}-x_0)$ and $\frac{1}{2}(x_{n_j}-y_{n_j})$ lie in $clB(0,r)$, $\{Tx_{n_j}\}$ is bounded by some constant c_0 and $||y_{n_j}-x_0|| \to 0$, it follows from (12) that as $j \to \infty$ we have the relation

$$|(Tx_{n_j},K(\tfrac{1}{2}(x_{n_j}-x_0))-K(\tfrac{1}{2}(x_{n_j}-y_{n_j})))| \le c_0\psi(\tfrac{1}{2}||y_{n_j}-x_0||) \to 0 .$$

In view of this and the fact that $(Tx_{n_j},K(x_{n_j}-y_{n_j})) \to 0$, the equality (11) implies that

$$(Tx_{n_j}, K(x_{n_j} - x_0)) \to 0 \quad \text{as} \quad i \to \infty .$$

This and the fact that $(Tx_0, K(x_{n_j} - x_0)) \to 0$ as $j \to \infty$ imply that

$$(Tx_{n_j} - Tx_0, K(x_{n_j} - x_0)) \to 0 \quad \text{with} \quad x_{n_j} \rightharpoonup x_0 \quad \text{as} \quad j \to \infty .$$

Since, by assumption, T satisfies condition (KS) it follows that $x_{n_j} \to x_0$ as $j \to \infty$. To show that $Tx_0 = g$ note first that, by the demicontinuity of T, $Tx_{n_j} \rightharpoonup Tx_0$ in Y. Now, for any $w \in R(K)$ there exists $x \in X$ such that $w = Kx$ and a sequence $z_n \in X_n$ such that $z_n \to x$ in X and $Kz_n \to Kx$ in Y^*. Thus, using the condition (b2), we see that for each $w \in R(K)$ we have the relation

$$\begin{aligned}(g - Tx_0, w) = (g - Tx_0, Kx) &= \lim_j (g - Tx_{n_j}, Kz_{n_j}) \\ &= \lim_j (Q_{n_j} g - Q_{n_j} Tx_{n_j}, Kz_{n_j}) = 0 .\end{aligned}$$

Since $R(K)$ is dense in Y^* and $(g - Tx_0, w) = 0$ for each $w \in R(K)$, it follows that $Tx_0 = g$. Hence T is A-proper w.r.t. Γ_a. Q.E.D.

Remark 6. We note that under the above conditions on K, the conclusion of Lemma 1 certainly holds if $T: X \to Y$ is required to satisfy the stronger condition (KS_+). In fact, it is not hard to show (see [5, 24]) that in the latter case if T_1 and T_2 satisfy condition (KS_+), then $T_t = tT_1 + (1-t)T_2$ is A-proper for each $t \in [0,1]$ or, equivalently, $T_1 + \mu T_2$ is A-proper w.r.t. Γ_a for each $\mu \geq 0$.

Remark 7. If X is reflexive and $T: X \to X^*$, then $Y = X^*$ and $Y^* = X$ and so in this case we may take $K = I$ and note that K thus chosen satisfies all the conditions of Lemma 1 if we take $Q_n = V_n^*$ and thus use the injective scheme Γ_d. In the latter case Lemma 1 and the convexity of the class (S_+) was first established by Browder [5] under the assumption that T is bounded and continuous. We add that condition (b4) used in [28] is analogous to the requirement that T be strongly quasibounded as defined by Browder and Hess [8]. We further add that as was shown in [28] and independently in [8], demicontinuous monotone mappings as well as pseudo-monotone mappings $T: X \to X^*$ as defined by Brezis (for separable reflexive spaces) are examples of mappings which satisfy condition (b4).

Remark 8. We note in passing that, in addition to the regularity theory for quasilinear elliptic equations, in his monograph [31] Skripnik developes also a degree theory, bifurcation theory, the approximation-solvability theory, and others for equations involving bounded demicontinuous mappings T from a real separable reflexive space X into X^* which satisfy condition (α) (i.e., T is such that if $x_j \rightharpoonup x$ in X and $\lim(Tx_j - Tx, x_j - x) \leq 0$, then $x_j \to x$ in X). It is not hard to show that if T is bounded and demicontinuous, then T satisfies condition (α) if and only if T satisfies condition (S_+). In view of Lemma 1 for $Y = X^*$ and $K = I$ and the preceeding remarks, the class of bounded demicontinuous maps which satisfy condition (α) studied in [31] form a convex subclass of A-proper mappings w.r.t. Γ_d. Consequently, a lot of results obtained in [31] follow from earlier results of Petryshyn, Browder-Petryshyn, Browder, Fitzpatrick, and others (see [27] for the references and the survey of various results).

For the benefit of the reader we recall

Definition 5. $T: X \to X^*$ is said to be

(1) monotone if $(Tx-Ty, x-y) \geq 0$ for all $x, y \in X$.

(2) pseudo-monotone if $x_j \rightharpoonup x$ in X and $\lim\sup(Tx_j, x_j - x) \leq 0$ imply that $(Tx, x-v) \leq \lim\inf(Tx_j, x_j - v)$ for all $v \in X$.

(3) semimonotone if there is $V: X \times X \to X^*$ such that $T(x) = V(x,x)$ for $x \in X$, $V(x,\cdot)$ is monotone and hemicontinuous and $V(\cdot,x)$ is completely continuous for each fixed $x \in X$.

(4) quasimonotone if $x_j \rightharpoonup x$ in X implies that $\lim\sup(Tx_j, x_j - x) \geq 0$.

Monotone maps were introduced independently by Vainberg-Kachurovsky and Zarantonello, and further studied by Minty, Browder, Kachurovsky, Rockafellar and others (see [5, 17] for references). The study of pseudo-monotone maps (somewhat differently defined) was initiated by Brezis [1] and those of semimonotone maps by Browder [3]. Hess [16] and Calvert-Webb [10] introduced independently the notion of quasi-monotone (or of type (P)) maps. Extending the result of Browder [5] for bounded pseudo-monotone maps, it was shown in [10, 16] that if X is reflexive and $T: X \to X^*$ is demicontinuous and quasi-monotone, then $T + J$ satisfies condition (S_+) for each $\mu > 0$ where (in view of the results of Kadec and Asplund) we may assume without loss of generality that X and X^* are locally uniformly convex and J is the normalized duality map of X into X^* given by $(Jx,x) = ||x||^2$ and $||Jx|| = ||x||$ for $x \in X$. It was shown by Fitzpatrick [15] that if T is either semi-monotone or locally bounded and pseudo-monotone, then T is demicontinuous and quasi-monotone.

We add in passing that maps $T: X \to Y$ which are K-monotone or of K-monotone type with $K: X \to Y^*$ not necessarily linear have been studied by Kato [18], Browder-DeFigueiredo [7], Browder [6], Petryshyn [23, 24], Dubinsky [12], Deimling [11], and others.

For the purpose of this paper we say, following [10, 16], that $T: X \to Y$ is <u>K-quasimonotone</u> if $x_j \rightharpoonup x$ in X implies that $\lim\sup(Tx_j, K(x_j - x)) \geq 0$.

<u>Lemma 2</u>. Suppose X is reflexive and $K: X \to Y^*$ satisfies conditions of Lemma 1. Suppose $T: X \to Y$ is a K-demicontinuous K-quasimonotone map which satisfies condition (b4) of Lemma 1. If $G: X \to Y$ is a bounded demicontinuous map which satisfies condition (KS_+), then $T_\mu = T + \mu G$ is A-proper w.r.t. Γ_d for each $\mu > 0$.

<u>Proof</u>: In view of Lemma 1 and Remark 6, it suffices to show that $T_\mu = T + \mu G$ satisfies condition (b4) and condition (KS_+) for each fixed $\mu > 0$.

Now, since G is bounded and T satisfies (b4), it follows easily that T_μ satisfies condition (b4) for each fixed $\mu > 0$. To show that T_μ satisfies condition (KS_+) let $x_j \rightharpoonup x$ in X and $\lim\sup(T_\mu x_j - T_\mu x, K(x_j - x)) \leq 0$. Since T is K-quasimonotone and $\mu > 0$, it follows that $\lim\sup(Gx_j - Gz, K(x_j - x)) \leq 0$. Hence $x_j \to x$ in X as $j \to \infty$ since G satisfies condition (KS_+). Q.E.D.

Now, in view of Lemmas 1 and 2, an immediate consequence of Theorem 4 is the following.

<u>Theorem 5</u>. Let X be reflexive and $K: X \to Y^*$ satisfy conditions of Lemma 1. Let $M_n: X_n \to Y_n$ satisfy condition (C2) with $K_n = Q_n^* K$. Suppose $T: X \to Y$ is a K-quasimonotone demicontinuous map such that T satisfies condition (++) and condition (b4) of Lemma 1. Suppose there is a bounded demicontinuous map $G: X \to Y$ which satisfies condition (KS_+). Suppose further that any one of the following two conditions

hold:

(i) there is $r_0 > 0$ such that T and G are odd on $X-B(0,r_0)$.

(ii) there is $r_0 > 0$ such that $(Tx,Kx) \geq 0$ for $||x|| \geq r_0$.

Then, if T satisfies condition (+), $T(X) = Y$.

In view of Remark 2, it is obvious that for separable spaces Theorem 5 extends and unifies a number of surjectivity results obtained earlier by other authors (using different methods) for maps T of monotone, J-monotone, and K-monotone type depending on the choice of Y, K and Γ (see [27] for some of the above special cases).

We conclude this section with some specific corollaries of Theorem 5 as they appeared in chronological order so as to show its relation to previous results for maps $T: X \to X^*$ of monotone type and for mappings $T: X \to X$ of J-monotone type. For the various consequences of Theorem 4 for the mappings of K-monotone type with $K \neq \Gamma$ or J see [24, 27].

To deduce the surjectivity results for maps $T: X \to X^*$ from Theorem 5 when X is separable and reflexive we set $Y = X^*$ and note that if we choose $K = I$, $\Gamma_a = \Gamma_d$, $K_n = V_n$, $G = J$ and define $M_n : X_n \to X_n^* \ (= Y_n)$ by $M_n(x) = \sum_{j=1}^{n} (f_j, x) f_j$, where $\{\psi_1, \ldots, \psi_n\}$ is a basis in X_n and $\{f_1, \ldots, f_n\}$ is the corresponding biorthogonal basis in X_n^* for each fixed integer $n \geq 1$, then all the conditions of Theorem 5 are satisfied provided that T satisfies the imposed conditions. In what follows we assume without loss of generality that X and X^* are locally uniformly convex.

Since a demicontinuous monotone map $T: X \to X^*$ satisfies condition (b4) (see [8, 28]), it follows from Lemma 2 that $T_\mu = T + \mu J$ is A-proper w.r.t. the injective scheme $\Gamma_d = \{X_n, V_n; X_n^*, V_n^*\}$. Hence our first corollary of Theorem 5 is the following result.

Corollary 1. If $T: X \to X^*$ is monotone demicontinuous and satisfies condition (+), then $T(X) = X^*$.

In case T is coercive the above result was proved independently by Browder [2] and Minty [21]. In its present form Corollary 1 was proved by Rockafellar [30] although under some additional conditions on X it was proved earlier by Browder [4].

Corollary 2. Let $T: X \to X^*$ be either semimonotone or pseudomonotone and bounded. Then, if T is coercive, $T(X) = X^*$.

In the semimonotone case Corollary 2 is due to Browder [3], while for bounded pseudomonotone maps it is essentially due to Brezis [1] (see also the results of [8, 13, 28]).

Corollary 3. Let $T: X \to X^*$ be pseudomonotone, bounded, and $(Tx,x) \geq (T(0),x) - c||x||$ for $x \in X$ and some $c \geq 0$. If, T is norm-coercive, then $T(X) = X^*$.

Corollary 3 was first deduced by the author in [25] from more general results on pseudo-A-proper mappings (see also [8] for an analogous result).

Corollary 4. Let $T: X \to X^*$ be demicontinuous pseudomonotone, bounded, and odd on $X - B(0,r_0)$. If T is also positively homogeneous of order $\alpha > 0$ and $0 \notin T(\partial B(0,1))$, then $T(X) = X^*$.

Corollary 4 was proved in [5] for T continuous. Under weaker continuity conditions it was established in [14, 26].

Now using the full force of Theorem 5 we see that the above corollaries can be unified and generalized into the following proposition which is also a special case of Theorem 5.

Proposition 1. Let $T: X \to X^*$ be a demicontinuous quasimonotone map which satisfies condition (++) and condition (b4) of Lemma 1 for $K = I$. Suppose further that T satisfies one of the following two conditions:

(i) there is $r_0 > 0$ such that T is odd on $X - B(0,r_0)$.

(ii) there is r_0 such that $T(x,x) \geq 0$ for $||x|| \geq r_0$.

Then, if T satisfies condition (+), $T(X) = X^*$.

In case T is bounded and coercive the above result was proved by Calvert and Webb [10]. In its present form it was proved by Fitzpatrick [15] without using condition (b4).

It should be added that the above corollaries and the proposition have been established by different methods and for spaces which need not be separable. Our purpose was to show that the A-proper mapping theory can be used to deduce rather general existence results for mappings which need not be A-proper or act from X to X^*.

To deduce the surjectivity results for maps $T: X \to X$ from Theorems 4 and 5 when X is a separable Banach space with a projectionally complete scheme $\Gamma_b = \{X_n, V_n; X_n, P_n\}$, we set $Y = X$ and note that if, for example, we choose $K = J: X \to X^*$, $K_n = P_n^* J$,

$M_n = I_n$ and $G = I$, then Theorem 4 yields the following new results.

Proposition 2. Suppose $T: X \to X$ satisfies condition (++) with $T_n: X_n \to X_n$ continuous and $T_\mu = T + \mu I$ A-proper w.r.t. Γ_b for each $\mu > 0$. Suppose further that one of the following conditions holds:

(i) there is $r_0 > 0$ such that T is odd on $X - B(0,r_0)$.

(ii) there is $r_0 > 0$ such that $(Tx,Jx) \geq 0$ for $||x|| \geq r_0$.

Then, if T satisfies condition (+), $T(X) = X$.

In [27] the writer surveyed various classes of maps $T: X \to X$ for which it has been shown that $T_\mu = T + \mu I$ is A-proper w.r.t. Γ_b for each $\mu > 0$. Consequently, Proposition 2 is applicable to these mappings and thus one can deduce from it a number of new as well as some known surjectivity results obtained earlier by various authors using different methods. See [27] for some special cases of Proposition 2.

It should be added that if X is reflexive and $K = J$ is weakly and strongly continuous, (which is always the case when $K: X \to X^*$ is linear and bounded, as is used in Section 2), then Theorem 5 shows that $T + \mu I$ is A-proper if $T: X \to X$ is K-quasimonotone and $\{Tx_j\}$ is bounded whenever $\{x_j\}$ and $\{(Tx_j, Kx_j)\}$, are bounded.

Section 2.

In this section we apply the results of Section 1 obtained for A-proper maps to the approximation-solvability of quasi-linear elliptic boundary value problems (BVP) of order 2 to obtain strong solutions in the Sobolev W_2^2 space. Note that our approach differs from that used by other authors and at the same time provides the convergence of the approximants in the W_2^2-norm.

Second order elliptic boundary value problems. Let $Q \subset R^n$ be a bounded domain with a sufficiently smooth boundary ∂Q so that the Sobolev Imbedding Theorem holds (see [4, 19]).

First BVP. Suppose L is a general elliptic operator defined on $D(L) = \mathring{W}_2^2 = W_2^2 \cap \mathring{W}_2^1$ by

$$L(u) = -\sum_{i,j=1}^{n} a_{ij}(x)u_{x_i x_j} + \sum_{i=1}^{n} a_i(x)u_{x_i} + a_0(x)u \tag{13}$$

with $a_{ij}(x) \in C^1(Q)$ and $a_i(x) \in C(Q)$ for $i = 0,1,\ldots,n$. The spaces $W_2^2(Q)$ and $W_2^1(Q)$ are the ordinary Sobolev spaces with $\overset{\circ}{W}{}_2^1$ denoting the completion in W_2^1 of the set $C_0^\infty(Q)$. We assume that L is strongly elliptic, i.e.,

$$\sum_{i,j=1}^{n} a_{ij}(x)\xi_i\xi_j \geq c_0\left(\sum_{i=1}^{n} |\xi_i|^2\right) \quad \text{for} \quad x \in \bar{Q}\ ,\ \{\xi_1,\ldots,\xi_n\} \in R^n \text{ and } c_0 > 0.$$

Our problem is to establish the strong approximation-solvability of the quasilinear equation on $\overset{\circ}{W}{}_2^2$ of the form.

$$Lu + c(x,u,u_{x_1},\ldots,u_{x_n}) = h(x) \quad (u \in \overset{\circ}{W}{}_2^2\ ,\ h \in L_2)\ , \tag{14}$$

where the function $c(x,u,u_{x_1},\ldots,u_{x_n})$ satisfies the following

Assumption (A): For each u in W_2^1 the map $u \mapsto Cu \equiv c(x,u,u_{x_1},\ldots,u_{x_n})$ yields a continuous mapping of W_2^1 into L_2 .

In what follows an element $u \in \overset{\circ}{W}{}_2^2$ satisfying Equation (14) a.e. will be called a strong solution of Equation (14).

To use the theory of A-proper mappings for the constructive solvability of Equation (14) in $\overset{\circ}{W}{}_2^2$ we first note that if we let K be a map defined on $D(K) = \overset{\circ}{W}{}_2^2$ by $Ku = -\Delta u$, then it is known that K is a selfadjoint positive definite map of $\overset{\circ}{W}{}_2^2$ onto L_2 and there exist constants $m_1 > 0$ and $m_2 > 0$ such that

$$m_1||Ku||^2 \leq ||u||_{2,2}^2 \leq m_2||Ku||^2 \quad \forall\, u \in \overset{\circ}{W}{}_2^2\ . \tag{15}$$

Let H_0 be the Hilbert space $\overset{\circ}{W}{}_2^2$ whose inner product and the equivalent norm are given by

$$[u,v] = (Ku,Kv)\ ,\ |u| = ||Ku|| \quad (u \in \overset{\circ}{W}{}_2^2)\ . \tag{16}$$

Let $\{\psi_i\} \subset H_0$ be linearly independent and complete in H_0 . Then $\{K\psi_i\}$ is linearly independent and complete in L^2 and so if $P_n : H_0 \to X_n$ and $Q_n : L_2 \to Y_n$ denote the orthogonal projections in H_0 and L_2 respectively with $X_n = sp\{\psi_1,\ldots,\psi_n\} \subset H_0$ and $Y_n = sp\{K\psi_1,\ldots,K\psi_n\} \subset L_2$, then $\Gamma_0 = \{X_n,P_n;\ Y_n,Q_n\}$ is projectionally complete for the pair (H_0,H) .

The projection method for Equation (14) consists in finding an $x_n \in X_n$ from the finite dimensional quasilinear equation

$$Q_nL(x_n) + Q_nC(x_n) = Q_n(h) \quad (x_n \in X_n\ ,\ Q_nh \in Y_n)\ . \tag{17}$$

It follows from (15) and our conditions on the coefficients of L that

L and K, considered as mappings from H_0 to L_2, are bounded. By the inequality of Sobolevsky [32] (see also [19]), there exist constants $\tau_1 > 0$ and $\tau_2 > 0$ such that

$$(Lu,Ku) \geq \tau_1 \int_Q \sum_{i,j=1}^{n} \left(\frac{\partial^2 u}{\partial x_i \partial x_j}\right)^2 dx - \tau_2 ||u||_{1,2}^2 \quad \text{for} \quad u \in H_0 .$$

It is easy to see that this inequality can be put in the form

$$(Lu,Ku) \geq \gamma_1 |u|^2 - \tau_2 [C_0 u,u] \quad \text{for} \quad u \in H_0 \quad \text{with} \quad \gamma_1 > 0 , \tag{18}$$

where $C_0 \in L(H_0,H_0)$ is a completely continuous map determined by the bilinear form $(u,v)_1$ on H_0 since, by the Sobolev Imbedding Theorem [4], the injection of $\overset{\circ}{W}{}_2^2$ into W_2^1 is compact. Hence, by the special case of Theorem 5 in [24] (i.e., Lemma 2.1H in [27]), $L: H_0 \to L_2$ is A-proper w.r.t. Γ_0. It follows from our construction of the scheme Γ_0 that since $Y_n = K(X_n)$ for each n, conditions (C1) and (C2) of Theorem 1 hold if we set $K_n = K|_{X_n}: X_n \to Y_n$ and take $M_n = K_n$ for each n.

Now, again from the Sobolev Imbedding Theorem and Assumption (A), it follows that if $\{u_n\}$ is a sequence in H_0 such that $u_n \rightharpoonup u_0$ in H_0, then $u_n \to u_0$ in W_2^1 and hence $C(u_n) \to C(u_0)$ in L_2. Hence $C: H_0 \to L_2$ is completely continuous and therefore compact since Hilbert spaces are reflexive. Consequently, $T = L+C: H_0 \to L_2$ is A-proper w.r.t. Γ_0 and, therefore, Theorem 1 implies the validity of the following new result for Equation (14).

<u>Theorem 6</u>. Let L be the strongly elliptic operator defined on H_0 by (13) and let C be the nonlinear operator satisfying the Assumption (A). Let Γ_0 be the scheme determined by $\{\psi_i\} \subset H_0$ and $\{K\psi_i\} \subset L_2$, where $Ku = -\Delta u$ for $u \in H_0$. Suppose that one of the following conditions hold:

(i) $c(x,-\xi_0,-\xi_1,\ldots,-\xi_n) = -c(x,\xi_0,\xi_1,\ldots,\xi_n)$ for all $(\xi_0,\ldots,\xi_n)$.

(ii) $(Tu,Ku) = (Lu,Ku) + (Cu,Ku) \geq 0$ for $u \in H_0$ with $|u| \geq r_0$ for some $r_0 > 0$.

If $T = L+C$ satisfies condition (+) (i.e., if $\{u_j\} \subset H_0$ is any sequence such that $||Tu_j - g|| \to 0$ for some g in L_2, then $\{|u_j|\}$ is bounded), then Equation (14) is feebly approximation-solvable in H_0 w.r.t. Γ_0 for each $f \in L_2$. Equation (14) is strongly approximation-solvable if it is uniquely solvable for a given $f \in L_2$ (i.e., in this case $|u_n - u_0| \to 0$ for some $u_0 \in H_0$, $T(u_0) = f$, and $||T(u_n)-f|| \to 0$ as $n \to \infty$).

Remark 9. In case C is odd (i.e., condition (i) holds), the conditions of Theorem 6 hold if condition (+) is replaced by any one of the conditions mentioned in Remark 2. This allows us a considerable freedom in imposing conditions on C (i.e. on the function c) so that one of the above conditions holds.

Remark 10. If C is not odd and L is such that $\tau_2 = 0$ in (18), then T satisfies condition (ii) and (+) if, for example, $(Cu,Ku) \geq -c_0|u|^\alpha$ for all $u \in H$ and some $c_0 > 0$ and $\alpha \in (0,2)$ since in that case T is K-coercive and thus in this case the conclusions of Theorem 6 follow.

It should be noted, however, that since L is an A-proper linear (and thus odd) map the assumption that L is one-to-one implies that $||Lu|| \geq \eta|u|$ for all $u \in H_0$ and some $\eta > 0$ and, in particular, that $||Lu|| \to \infty$ as $|u| \to \infty$. Hence applying Theorem 3 to the approximation-solvability of Equation (14) we get the following.

Theorem 7. If L is one-to-one, C satisfies Assumption (A), and $||Lu|| - ||Cu|| \to \infty$ as $|u| \to \infty$, then Equation (14) is feebly approximation-solvable for each $f \in L_2$. Equation (14) is strongly approximation-solvable if it is uniquely solvable.

Proof: It is easy to show that an A-proper linear map $L: H_0 \to L_2$ satisfies condition (+) if and only if L is one-to-one. Since $L + tC$ is A-proper w.r.t. Γ_0 for each $t \in [0,1]$, in view of Theorem 3, to prove Theorem 7 it suffices to verify condition (C4) for $T_t = L + tC$. Let h be an arbitrary but fixed element in L_2 and suppose that $Lu + tCu = h$ holds for some $u \in H_0$ and $t \in [0,1]$. Since $||h|| = ||Lu+tCu|| \geq ||Lu|| - ||Cu||$, it follows from our condition that $||Lu|| - ||Cu|| \to \infty$ whenever $||u|| \to \infty$, that there exists $c(||h||) > 0$ such that $||u|| \leq c(||h||)$. Hence the conclusions of Theorem 7 follow from Theorem 3. Q.E.D.

In view of Remark 4, the following corollary to Theorem 7 appears to be useful from the practical point of view since the corresponding conditions are easily verifiable although more restrictive.

Corollary 5. Let L be one-to-one and C satisfy Assumption (A). If $C: H_0 \to L_2$ is quasibounded with its quasinorm $|C| < \eta$ and, in particular, if C is asymptotically zero, then the conclusions of Theorem 7 hold.

Proof: Since C is quasibounded, it follows that to each $\varepsilon > 0$ there exists $r > 0$ such that $||Cu|| \leq (|C|+\varepsilon)|u|$ for all $u \in H_0$ with $|u| \geq r$. Hence, since $\eta > |C|$, we can choose an $\varepsilon > 0$ so that $\eta-|C|-\varepsilon > 0$ and obtain the relation

$$||Lu|| - ||Cu|| \geq (\eta-|C|-\varepsilon)|u| \to \infty \quad \text{as} \quad |u| \to \infty . \tag{19}$$

In view of Theorem 7, (19) implies the validity of Corollary 5. Note that (19) is always true if C is asymptotically zero since in this case $|C| = 0$. Q.E.D.

Remark 11. Using the Sobolev Imbedding Theorem and the Vainberg Theorem for Nimytzkii operators (see [33]) one can show that Assumption (A) is implied by the following.

Condition I. The function $c(x,u,u_{x_1},\ldots,u_{x_n})$ is measurable in $x \in Q$ for fixed $(u,u_{x_1},\ldots,u_{x_n})$ and continuous in $(u,u_{x_1},\ldots,u_{x_n})$ for almost all $x \in Q$. Furthermore, there exist $\psi(x) \in L_2$ and constants $b_1 > 0$ and $b_2 > 0$ such that

$$|c(x,u,u_{x_1},\ldots,u_{x_n})| \leq \psi(x) + b_1|u|^{\gamma} + b_2\left(\sum_{i=1}^{n} |u_{x_i}|\right) ,$$

where $\gamma = n/(n-2)$ with $n > 2$.

Third BVP. Assuming that ∂Q is sufficiently smooth (see [19]) we let $W^2_{2,0}$ denote the closed subspace of functions in W^2_2 satisfying the boundary condition

$$\frac{\partial u}{\partial \ell} + \delta u = 0 \quad \text{for} \quad x \in \partial Q , \tag{20}$$

where $\frac{\partial}{\partial \ell}$ denotes the derivative along the given outward pointing direction on ∂Q and where $\delta(s)$ is some given bounded function on ∂Q.

Suppose that L is the strongly elliptic operator with sufficiently smooth coefficients which is defined on $D(L) = W^2_{2,0}$ and which is of the form

$$Lu = \sum_{i,j=1}^{n} \frac{\partial}{\partial x_i}\left(a_{ij}(x)\frac{\partial u}{\partial x_j}\right) - \lambda_0 u + \sum_{i=1}^{n} a_i(x)\frac{\partial u}{\partial x_i} + au , \qquad (21)$$

where λ_0 is a number so chosen (see [19]) that the operator

$$L_0(u) = \sum_{i,j}^{n} \frac{\partial}{\partial x_i}\left(a_{ij}(x)\frac{\partial u}{\partial x_j}\right) - \lambda_0 u$$

determines a bijective mapping of $W^2_{2,0}$ onto L_2 so that for some constant $m_1 > 0$

$$m_1 ||u||^2_{2,2} \le ||L_0 u|| \quad \text{for all } u \text{ in } W^2_{2,0} .$$

Our problem is to establish the approximation-solvability of the quasilinear equation in $W^2_{2,0}$ of the form

$$Lu + c(x,u,u_{x_1},\ldots,u_{x_n}) = h(x) \quad (u \in W^2_{2,0} , h \in L_2) , \qquad (23)$$

where the function $c(x,u,u_{x_1},\ldots,u_{x_n})$ satisfies the Assumption (A).

In view of (23), we can let H^0 be the Hilbert space $W^2_{2,0}$ whose inner product and the equivalent norm are given by $[u,v]_0 = (Ku,Kv)$, $|u|_0 = ||Ku||$ with $K = L_0$ and $u \in W^2_{2,0}$. Let $\{\psi_i\} \in H^0$ be linearly independent and complete in H^0 . Then $\{K\psi_i\}$ is linearly independent and complete in L_2 . As before, the scheme $\Gamma^0 = \{X_n,P_n; Y_n,Q_n\}$ is projectionally complete for the pair (H^0,L_2) determined by $\{\psi_i\} \subset H^0$ and $\{K\psi_i\} \equiv \{L_0\psi_i\} \subset L_2$.

If we set $L_1(u) = \sum_{i=1}^{n} a_i(x)u_{x_i} + au$ and note that $|(L_1u,L_0u)| \le b||u||_{W_2}\cdot||u||_{W_2^2}$ for all $u \in H^0$, then by Sobolev imbedding Theorem there is a compact map $C^0 \in L(H^0,H^0)$ such that $(L_1u,L_0v) = [C_0u,v]_0$ for all $u,v \in H^0$. Since $(Lu,Ku) = |u|^2_0 + [C^0u,u]_0$ for all $u \in H^0$, it follows from Lemma 2.1H in [27] that $L: H^0 \to L_2$ is bounded A-proper w.r.t. Γ^0 . The same argument as before shows that $C: H^0 \to L_2$ defined by $C(u) = c(x,u,u_{x_1},\ldots,u_{x_n})$ is compact.

The above discussion shows that, using the A-proper mapping theorems established in Section 1, one can make the same assertions concerning the approximation-solvability of Equation (23) in $H^0 = W^2_{2,0}$ corresponding to the Third BVP as one did about the Equation (14) in $H_0 = \mathring{W}^2_2$. We summarize this in the following result.

Theorem 8. Let L be the strongly elliptic operator defined on H^0 by (21) and let C be the nonlinear operator satisfying Assumption (A). Let Γ^0 be the scheme determined by $\{\psi_i\} \subset H^0$ $(= W^2_{2,0})$ and $\{K\psi_i\} \subset L_2$, where $Ku = L_0 u$ for $u \in W^2_{2,0}$ and λ_0 is so chosen that L_0 determines a bounded bijective mapping of $W^2_{2,0}$ onto L_2 for which (22) holds. Then Equation (23) is feebly or strongly approximation-solvable in H^0 w.r.t. Γ^0 provided that the operators L and C satisfy conditions analogous to those imposed in Theorems 6 and 7 and Corollary 5.

---

* Supported in part by the NSF Grant GP-20228 and in part by the Research Council of Rutgers University while the author was on the faculty research fellowship during the year 1974-75.

+ The results contained in this paper were presented by the author at the "Seminar on Fixed Point Theory and its Applications" at the Dalhousie University, Halifax, N.S., Canada, June 9-12, 1975.

[1] See the succeeding sections for the precise definitions and assertions mentioned in the Introduction.

References

[1] Brezis, H., "Équations et inéquations non-lineaires dans les espaces en dualite." Ann. Inst. Fourier, Grenoble, 18 (1968), 115-175.

[2] Browder, F.E., "Nonlinear elliptic boundary value problems." Bull. AMS, 19 (1963), 862-874.

[3] __________, "Mapping theorems for noncompact nonlinear operators." Proc. Nat. Acad. Sci. USA, 54 (1965), 337-342.

[4] __________, "Existence theorems for nonlinear partial differential equations." Proc. Symp. Pure Math. (Berkeley) 16 (1968), 1-60.

[5] __________, "Nonlinear operators and nonlinear equations of evolution in Banach spaces." Proc. of Symp. on Nonlinear Analysis, AMS (to appear).

[6] __________, "Normal solvability and ϕ-accretive mappings of Banach spaces." Bull. Amer. Math. Soc. 78 (1972), 186-192.

[7] Browder, F.E. and D.G. de Figueiredo, "J-monotone nonlinear operators in Banach spaces." Nederl. Akad. Wetensch. Proc. Ser. A 69 = Indag. Math. 28 (1966), 671-676.

[8] Browder, F.E. and P. Hess, "Nonlinear mappings of monotone type in Banach spaces." J. Funct. Anal., 11 (1972), 251-294.

[9] Browder, F.E. and W.V. Petryshyn, "Approximation methods and the generalized topological degree for nonlinear maps in Banach spaces." J. Functional Anal., 3 (1969), 217-245.

[10] Calvert, B., and J.R.L. Webb, "An existence theorem for quasi-monotone operators." Rend. Accad. Naz. Dei Lincei, 8 (1971), 362-368.

[11] Deimling, K., "Zeros of accretive operators." Manuscripta Math., 13 (1974), 365-374.

[12] Dubinsky, Yu. A., "On some noncoercive nonlinear equations." Mat. Sb., 87 (1972), 315-323.

[13] de Figueiredo, D.G., "An existence theorem for pseudomonotone operator equations in Banach spaces." J. Math. Anal. Appl.

[14] Fitzpatrick, P.M., "A generalized degree for the uniform limit of A-proper mappings." J. Math. Anal. Appl., 35 (1971), 536-552.

[15] ________, "Surjectivity results for nonlinear mappings from a Banach space to its dual." Math. Ann., 204 (1973), 177-188.

[16] Hess, P., "On nonlinear mappings of monotone type homotopic to odd operators." J. Funct. Anal., 11 (1972), 138-167.

[17] Kachurovsky, R.I., "Nonlinear monotone operators in Banach spaces." Mat. Nauk, 23 (1968), 121-168.

[18] Kato, T., "Demicontinuity, hemicontinuity and monotonicity II." Bull. Amer. Math. Soc., 73 (1967), 886-889.

[19] Ladyzenskaya, O.L. and N.N. Uralceva, "Linear and quasilinear equations of elliptic type." "Nauka," Moskow (1973).

[20] Leray, J. and J.L. Lions, "Quelques résultats de Visik sur les problemes elliptiques non linéaries par les méthods de Minty-Browder." Bull. Soc. Math. France, 93 (1965), 97-107.

[21] Minty, G.J., "On a "monotonicity" method for the solution of non-linear equations in Banach spaces." Proc. Nat. Acad. Sci. USA, 50 (1963), 1038-1041.

[22] Petryshyn, W.V., "On the approximation-solvability of nonlinear equations." Math. Ann., 177 (1968) 156-164.

[23] ________, "Nonlinear equations involving noncompact operators, Proc. Symp. Pure Math., Vol. 18, part II, AMS. Providence, RI (1970), 206-233.

[24] ________, "Antipodes theorem for A-proper mappings and its application to mappings of the modified type (S) or (S+) and to mappings with the pm-property." J. Funct. Anal., 7 (1971), 165-211.

[25] ________, "On existence theorems for nonlinear equations involving noncompact operators." Proc. Nat. Acad. Sci. USA, 67 (1970), 326-330.

[26] ________, "Surjectivity theorems for odd maps of A-proper type." Math. Ann., 192 (1971), 155-172.

[27] ________, "On the approximation-solvability of equations involving A-proper and pseudo-A-proper mappings." Bull. Amer. Math. Soc., 81 (1975), 223-312.

[28] Petryshyn, W.V. and P.M. Fitzpatrick, "New existence theorems for nonlinear equations of Hammerstein type." Trans. Amer. Math. Soc., 160 (1971), 39-63.

[29] ________, "On 1-set and 1-ball contractions with application to perturbation problems for nonlinear bijective maps and linear Fredholm maps." Boll. UMI, 417 (1973), 102-124.

[30] Rockafellar, R.T., "Local boundedness of nonlinear maximal monotone operators." Mich. Math. J., (1969), 397-407.

[31] Skripnik, I.V., "Nonlinear elliptic equations of higher order." "Naukova Dumka," Kiev, 1973.

[32] Sobolevsky, P.E., "On equations with operators forming an acute angle." Dokl. Akad. Nauk SSSR, 116 (1957), 254-257.

[33] Vainberg, M.M., "Variational method and the method of monotone operators in the theory of nonlinear equations." Wiley (1974), New York.

[34] Webb, J.R.L., "A fixed point theorem and applications to functional equations in Banach spaces." Boll. UMI., 4 (1971), 775-788.

SOME RESULTS IN THE FIXED POINT THEORY OF NONEXPANSIVE MAPPINGS AND GENERALIZED CONTRACTIONS

J. Reinermann and R. Schöneberg

Introduction

It is the aim of this paper to give some new results on fixed points by a "Čebyšev-center method" applied to bounded sequences in normed linear spaces. This method and related concepts have been extensively studied by several authors (see [1], [4], [8], [9], [11], [21], [27]) and have had important applications in the fixed point theory of nonexpansive mappings (see [2], [8], [9], [20], [27]). By refining the argument we get fixed point theorems under boundary conditions of "inwardness type" (see [5]) for nonexpansive mappings and generalized contractions (in the sense of W.A. Kirk [15], [16], [17]) having domains, which are not necessarily convex. Some of these results are of constructive type (Picard-iteration, Toeplitz-iteration (see [28])). We get also partial solutions of some well-known fixed point problems for mappings being the sum of a generalized contraction and a compact mapping both defined on a weakly compact subset of a Banach-space.

For a normed linear space (n.l.s.) $(E,\|\ \|)$ and a subset X of E we let $\bar{X}$, $int(X)$ and $co(X)$ respectively denote the closure of X , the interior of X and the convex hull of X . The boundary of X (i.e. $\bar{X}\setminus int(X)$) is denoted by ∂X . If $f: X \to E$ is a map, **Fix(f)** denotes the fixed point set of f . For $y \in E$ and a nonempty subset M of E we let $d(y,M)$ denote the distance from y to M .

<u>Definition 1</u>. Let $(E,\|\ \|)$ be a n.l.s. For a bounded sequence $(u_n)_{n\in\mathbb{N}} \in E^{\mathbb{N}}$ we define the map $R((u_n)_{n\in\mathbb{N}}): E \to \mathbb{R}^+$ by

$$R((u_n)_{n\in\mathbb{N}})(x) := \overline{\lim}(\|u_n - x\|)_{n\in\mathbb{N}} \quad .$$

If X is a nonvoid subset of E we call

$$C(X,(u_n)_{n\in\mathbb{N}}) := \{x \mid x\in X \wedge R((u_n)_{n\in\mathbb{N}})(x) = \inf_{y\in X} R((u_n)_{n\in\mathbb{N}})(y)\}$$

<u>the Cebysev-center of $(u_n)_{n\in\mathbb{N}}$ with respect to X.</u>

<u>Definition 2</u>. Let $(E,\|\ \|)$ be a n.l.s., $\emptyset \neq X \subset E$ and $f\colon X \to E$.

(i) f is said to be <u>nonexpansive</u>: $\iff \forall_{x,y\in X} \ \|f(x)-f(y)\| \leq \|x-y\|$.

(ii) f is said to be a <u>generalized contraction</u>: $\iff$

$$:\iff \exists_{\alpha\colon X\to[0,1)} \ \forall_{x,y\in X} \ \|f(x)-f(y)\| \leq \alpha(x)\|x-y\| \ .$$

The main properties of Cebysev-centers we need here, are summarized in:

<u>Lemma 1</u>. Let $(E,\|\ \|)$ be a n.l.s., $\emptyset \neq X \subset E$, $f\colon X \to E$ and $(u_n)_{n\in\mathbb{N}} \in E^{\mathbb{N}}$ be a bounded sequence. Then

(i) Either $C(X,(u_n)_{n\in\mathbb{N}}) \subset C(E,(u_n)_{n\in\mathbb{N}})$ or $C(X,(u_n)_{n\in\mathbb{N}}) \subset \partial X$.

(ii) X being boundedly weakly compact (i.e. the intersection of X with every closed ball about the origin is weakly compact) implies $C(X,(u_n)_{u\in\mathbb{N}}) \neq \emptyset$.

(iii) If $u_n \in X$ for $n \in \mathbb{N}$ and there is $k \in \mathbb{Z}^+$ such that

$$\lim(d(u_{n+k},\mathrm{co}(\{f(u_v) \mid v\in\mathbb{N}\wedge v\geq n\})))_{n\in\mathbb{N}} = 0$$

then for every $z \in X$ and $\lambda \in [0,1]$ such that

$$\forall_{y\in X} \ \|f(y)-f(z)\| \leq \lambda\|y-z\|$$

we have

$$R((u_n)_{n\in\mathbb{N}})(f(z)) \leq \lambda R((u_n)_{n\in\mathbb{N}})(z) \ .$$

(iv) $(E,\|\ \|)$ being uniformly convex in every direction (u.c.e.d.) (i.e. $\forall_{\substack{z\in E\\ \|z\|=1}} \ \forall_{M,\Sigma>0} \ \exists_{\delta\in(0,1)} \ \forall_{x,y\in E} \ \forall_{t\in\mathbb{R}} \ [\|x\|,\|y\| \leq M \wedge$

$\wedge x-y = tz \wedge |t| \geq \Sigma \Rightarrow \frac{1}{2}\|x+y\| \leq (1-\delta)\max\{\|x\|,\|y\|\}]$,

and X being convex implies $\mathrm{card}(C(X,(u_n)_{n\in\mathbb{N}})) \leq 1$.

<u>Proof</u>. (i): If $C(X,(u_n)_{n\in\mathbb{N}})$ is not a subset of $C(E,(u_n)_{n\in\mathbb{N}})$ there is $v \in E$ such that $R((u_n)_{n\in\mathbb{N}})(v) < \delta := \inf_{y\in X} R((u_n)_{n\in\mathbb{N}})(y)$.

Let $x \in C(X,(u_n)_{n\in\mathbb{N}})$ and suppose $x \in \mathrm{int}(X)$. Choosing $\lambda \in (0,1)$ such that $\lambda x + (1-\lambda)v \in X$ we immediately find

$$\rho \leq R((u_n)_{n\in\mathbb{N}})(\lambda x+(1-\lambda)v) \leq \lambda\rho + (1-\lambda)R((u_n)_{n\in\mathbb{N}})(v)$$

a contradiction to $R((u_n)_{n\in\mathbb{N}})(v) < \rho$. Hence $x \in \partial X$.

<u>(ii)</u>: For each $\eta \in \mathbb{R}$ the preimage $R((u_n)_{n\in\mathbb{N}})^{-1}[(-\infty,\eta]]$ is clearly convex, closed and bounded, hence $X \cap R((u_n)_{n\in\mathbb{N}})^{-1}[(-\infty,\eta]]$ is weakly compact. Letting r denote $\inf\limits_{y\in X} R((u_n)_{n\in\mathbb{N}})(y)$ this implies $X\cap R((u_n)_{n\in\mathbb{N}})^{-1}[(-\infty,r]] = \bigcap\limits_{m\in\mathbb{N}} (X\cap R((u_n)_{n\in\mathbb{N}})^{-1}[(-\infty,r+\frac{1}{m}]]) \neq \emptyset$.
Because of $C(X,(u_n)_{n\in\mathbb{N}}) = X\cap R((u_n)_{n\in\mathbb{N}})^{-1}[(-\infty,r]]$ we are done.

<u>(iii)</u>: Let $\varepsilon > 0$. By assumption there is $n_0 \in \mathbb{N}$ such that for $n \geq n_0$, $d(u_{n+k},\mathrm{co}(\{f(u_\nu)\,|\,\nu\in\mathbb{N}\wedge\nu\geq n\})) \leq \varepsilon$ and $||u_n-z|| \leq R((u_n)_{n\in\mathbb{N}})(z) + \varepsilon$. This yields for $n \geq n_0$: $||u_{n+k}-f(z)|| \leq 2\varepsilon + \lambda(R((u_n)_{n\in\mathbb{N}})(z) + \varepsilon)$. Hence $R((u_n)_{n\in\mathbb{N}})(f(z)) \leq 3\varepsilon + \lambda R((u_n)_{n\in\mathbb{N}})(z)$. Letting ε tend to zero, the conclusion follows.

<u>(iv)</u>: Let $x,y \in C(X,(u_n)_{n\in\mathbb{N}})$ and assume $x \neq y$. Defining $\varepsilon := t := ||x-y||$, $z := t^{-1}(y-x)$ and $M := ||x||+||y||+ \sup\limits_{n\in\mathbb{N}}||u_n||$ there is $\delta \in (0,1)$ such that for $n \in \mathbb{N}$

$$||u_n - \tfrac{1}{2}(x+y)|| = \tfrac{1}{2}||(u_n-x)+(u_n-y)|| \leq (1-\delta)\max\{||u_n-x||,||u_n-y||\}$$

which implies $R((u_n)_{n\in\mathbb{N}})(\frac{1}{2}(x+y)) \leq (1-\delta)\inf\limits_{w\in X} R((u_n)_{n\in\mathbb{N}})(w)$, a contradiction to $x \neq y$.

Q.E.D.

<u>Remark 1</u>. For further properties of Cebysev-centers we refer to [1], [4], [11] and [22].

<u>Theorem 1</u>. Let $(E,||\ ||)$ be a n.l.s. which is u.c.e.d. Let $X \subset E$ be weakly compact and $f\colon X \to E$ be nonexpansive such that

$$\forall_{x\in\partial X}\ \exists_{\lambda\in(0,1)}\ \lambda x+(1-\lambda)f(x)\in X \tag{1.1}$$

$$\bigcap_{n\in\mathbb{N}} \mathrm{dom}(f^n) \neq \emptyset \tag{1.2}$$

Then $\mathrm{Fix}(f) \neq \emptyset$.

<u>Proof</u>. We remark first that $(E,||\ ||)$ being u.c.e.d. implies

$$\underset{\substack{z\in E\\ ||z||=1}}{\forall}\ \underset{\lambda\in(0,1)}{\forall}\ \underset{M,\varepsilon>0}{\forall}\ \underset{\delta\in(0,1)}{\exists}\ \underset{x,y\in E}{\forall}\ \underset{t\in\mathbb{R}}{\forall}\ [||x||,||y||\le M\wedge x-y=tz\ \ |t|\ge\varepsilon \Longrightarrow$$
$$\Longrightarrow ||\lambda x+(1-\lambda)y|| \le (1-\delta)\max\{||x||,||y||\}]$$

as is easily seen by the inequality

$$||\lambda x+(1-\lambda)y|| \le 2\lambda\ \frac{1}{2}||x+y||+(1-2\lambda)\max\{||x||,||y||\}$$

where $x,y \in E$ and $\lambda \in (0,\frac{1}{2}]$ (without loss of generality). Choose $x \in \bigcap_{n\in\mathbb{N}} \mathrm{dom}(f^n)$ and then fix $y \in C(X,(f^n(x))_{n\in\mathbb{N}})$. Assume now $f(y) \neq y$. By (1.1) there is $\lambda \in (0,1)$ such that $v:= \lambda y+(1-\lambda)f(y)\in X$. If we set $\varepsilon:= ||y-f(y))||$, $z:= \varepsilon^{-1}(y-f(y))$ and $M:= \mathrm{diam}(f[X]\cup X)$, there is by the remark above a $\delta \in (0,1)$ such that for $n \in \mathbb{N}$

$$||v-f^{n+1}(x)|| = ||\lambda(y-f^{n+1}(x))+(1-\lambda)(f(y)-f^{n+1}(x))||$$
$$\le (1-\delta)\max\{||y-f^{n+1}(x)||,||y-f^n(x)||\}\ .$$

This implies $R((f^n(x))_{n\in\mathbb{N}})(v) \le (1-\delta)\inf_{w\in X} R((f^n(x))_{n\in\mathbb{N}})(w)$, a contradiction to $f(y) \neq y$.

Q.E.D.

If we strengthen condition (1.1) to " $f[\partial X]\subset\mathrm{int}(X)$ ", condition (1.2) is unnecessary in theorem 1. This is the content of

<u>Theorem 2</u>. Let $(E,|| \ ||)$ be a n.l.s. which is u.c.e.d. Let X be a nonvoid weakly compact subset of E and $f: X \to E$ be non-expansive such that $f[\partial X]\subset\mathrm{int}(X)$. Then $\mathrm{Fix}(f) \neq \emptyset$.

<u>Proof</u>. Let $u_0 \in X$. A simple observation yields the existence of a sequence $(u_n)_{n\in\mathbb{N}} \in X^{\mathbb{N}}$ such that

$$\underset{n\in\mathbb{N}}{\forall}\ [(f(u_{n-1})\in X\Rightarrow u_n=f(u_{n-1}))\wedge(f(u_{n-1})\notin X\Rightarrow u_n\in\partial X\cap co(\{u_{n-1},f(u_{n-1})\}))]\ .$$

It is easy to see, that we have for $n \in \mathbb{Z}^+$: $u_{n+2} \in co(\{f(u_n,f(u_{n+1})\})$ which yields - by lemma 1 (iii) - for every $z\in X$: $R((u_n)_{n\in\mathbb{N}})(f(z)) \le$ $\le R((u_n)_{n\in\mathbb{N}})(z)$. This implies - observing $f[\partial X]\subset\mathrm{int}(X)$ - that $C(X,(u_n)_{n\in\mathbb{N}})$ is not a subset of ∂X. Therefore by lemma 1 (i) $C(X,(u_n)_{n\in\mathbb{N}})\subset C(E,(u_n)_{n\in\mathbb{N}})$. Choose now $x \in C(X,(u_n)_{n\in\mathbb{N}})$. We have shown, that $\{x,f(x)\}\subset C(E,(u_n)_{n\in\mathbb{N}})$ and hence $f(x) = x$ (by lemma 1 (iv)).

Q.E.D.

Remark 2. The hypothesis " $f[\partial X]\subset \mathrm{int}(X)$ " cannot be weakened to " $f[\partial X]\subset X$ " as is shown by the simple example: $(E,\|\ \|):=(\mathbb{R},|\ |)$, $X:=[-2,-1]\cup[1,2]$ and $f(x):=-x$.

If f is a generalized contraction, the uniform convexity assumptions can be omitted in theorems 1 and 2. This is proved in theorems 3 and 4 below.

Theorem 3. Let $(E,\|\ \|)$ be a n.l.s., $X\subset E$ be boundedly weakly compact and $f\colon X\to E$ be a generalized contraction such that

$$\underset{x\in\partial X}{\forall}\ \underset{\lambda\in[0,1]}{\exists}\ \lambda x+(1-\lambda)f(x)\in X \tag{3.1}$$

$$\bigcap_{n\in\mathbb{N}}\mathrm{dom}(f^n)\neq\emptyset \tag{3.2}$$

Then there is a unique $x\in X$ with $f(x)=x$ and for $y\in\bigcap_{n\in\mathbb{N}}\mathrm{dom}(f^n)$ we have $\lim(f^n(y))_{n\in\mathbb{N}}=x$ (strongly).

Proof. Let $y\in\bigcap_{n\in\mathbb{N}}\mathrm{dom}(f^n)$. Because for $n\in\mathbb{N}$ the inequality $\|f^n(y)-y\|\leq(1-\alpha(y))^{-1}\|f(y)-y\|$ is satisfied, the sequence $(f^n(y))_{n\in\mathbb{N}}$ is bounded and therefore - by lemma 1 (ii) - $C(X,(f^n(y))_{n\in\mathbb{N}})\neq\emptyset$. Choose $x\in C(X,(f^n(y))_{n\in\mathbb{N}})$ and then $\lambda\in[0,1)$ such that $\lambda x+(1-\lambda)f(x)\in X$. We immediately find - using lemma 1 (iii) -

$$\begin{aligned} R((f^n(y))_{n\in\mathbb{N}})(x) &\leq R((f^n(y))_{n\in\mathbb{N}})(\lambda x+(1-\lambda)f(x))\\ &\leq \lambda R((f^n(y))_{n\in\mathbb{N}})(x)+(1-\lambda)R((f^n(y))_{n\in\mathbb{N}})(f(x))\\ &\leq [\lambda+(1-\lambda)\alpha(x)]R((f^n(y))_{n\in\mathbb{N}})(x)\ . \end{aligned}$$

Hence $R((f^n(y))_{n\in\mathbb{N}})(x)=0$ i.e. $\lim(f^n(y))_{n\in\mathbb{N}}=x$ (strongly) and $f(x)=x$. The uniqueness of the fixed point is trivial.

Q.E.D.

Theorem 4. Let $(E,\|\ \|)$ be a n.l.s., $\emptyset\neq X\subset E$ be boundedly weakly compact and $f\colon X\to E$ be a generalized contraction such that $f[\partial X]\subset X$. Then there is exactly one fixed point of f and for $y\in\bigcap_{n\in\mathbb{N}}\mathrm{dom}(f^n)$ the successive approximants $(f^n(y))_{n\in\mathbb{N}}$ converge strongly to this fixed point.

Proof. Let $u_0 \in X$ and choose a sequence $(u_n)_{n\in \mathbb{N}} \in X^{\mathbb{N}}$ as in the proof of theorem 2. Using $u_{n+2} \in co(\{f(u_n), f(u_{n+1})\})$ we immediately find the estimation $||u_n - u_0|| \leq (1-\alpha(u_0))^{-1}||f(u_0)-u_0||$. Hence $(u_n)_{n\in \mathbb{N}}$ is bounded, which yields $C(X,(u_n)_{n\in \mathbb{N}}) \neq \emptyset$. Let $z \in C(X,(u_n)_{n\in \mathbb{N}})$ be fixed. By lemma 1 (iii) we get $R(u_n)_{n\in \mathbb{N}})f(z)) \leq \alpha(z)R((u_n)_{n\in \mathbb{N}})(z)$. This yields - observing lemma 1 (i) and $f[\partial X] \subset X - R((u_n)_{n\in \mathbb{N}})(z) = 0$. Hence $R((u_n)_{n\in \mathbb{N}})(f(z)) = 0$, too, and therefore $f(z) = z$. The remaining part of the assertion is evident. Q.E.D.

Remark 3 (1). Theorems 3 and 4 improve corresponding results of S. Reich [27], W.A. Kirk [15], [16], [17] and others. If X is assumed to be convex in theorem 3 the condition (3.2) is unnecessary (see [27]). It is easy to see, that for a continuous self-map f of X we need only assume that an iterate of f is a generalized contraction in theorem 4.

(2). Let $(E,|| \, ||)$ be a n.l.s., $X \subset E$ be boundedly weakly compact and convex, $x_1 \in X$, $f\colon X \to X$ be a generalized contraction and let a TOEPLITZ-iteration procedure be defined by

$$x_{n+1} := (1-c_n)x_n + c_n f(x_n) \qquad ((c_n)_{n\in \mathbb{N}} \in [0,1]^{\mathbb{N}} \text{ with } \sum_{n=1}^{\infty} c_n = \infty)$$

then $(x_n)_{n\in \mathbb{N}}$ converges strongly to the unique fixed point of f. The proof is a consequence of the inequality

$$||x_{n+1} - y|| \leq \prod_{v=1}^{n} [1-c_v(1-\alpha(y))]\,||x_1 - y||$$

where y denotes the unique fixed point of f and which is easily proved by induction.

(3). Theorems 3 and 4 are partially generalized in [22]. If $(E,|| \, ||)$ is a reflexive Banach-space, theorem 4 states, that every self-mapping f of E, which is a generalized contraction, has exactly one fixed point (hence $I-f$ is bijective). In order to generalize this, we prove

Theorem 5. Let $(E,|| \, ||)$ be a Banach-space which is center-point complete (i.e. for every bounded sequence $(u_n)_{n\in \mathbb{N}} \in E^{\mathbb{N}}$ we have $C(E,(u_n)_{n\in \mathbb{N}}) \neq \emptyset$) and $f\colon E \to E$ be a generalized contraction. Then there is a unique $x \in E$ such that $f(x) = x$ and for every $y \in E$ we have $\lim(f^n(y))_{n\in \mathbb{N}} = x$ (strongly).

Proof. Let $y \in E$. Since $(f^n(y))_{n\in \mathbb{N}}$ is bounded (see the proof of theorem 3) we have $C(E,(f^n(y))_{n\in \mathbb{N}}) \neq \emptyset$. Fix $x \in C(E,(f^n(y))_{n\in \mathbb{N}})$. Using lemma 1 (iii) we get

$$R((f^n(y))_{n\in \mathbb{N}})(x) \leq R((f^n(y))_{n\in \mathbb{N}})(f(x)) \leq \alpha(x)R((f^n(y))_{n\in \mathbb{N}})(x) .$$

Hence $R((f^n(y))_{n\in \mathbb{N}})(x) = 0$ i.e. $\lim(f^n(y))_{n\in \mathbb{N}} = x$ (strongly) and $f(x) = x$.

Q.E.D.

Remark 4 (1). Every center-point complete normed linear space is a Banach-space, but the converse isn't true in general (see [22]).

(2). Every reflexive Banach-space is center-point complete (see lemma 1 (ii)).

(3). Using an argument due to J.R. Calder et. al. [4] it was proved in [1], that every Banach-space which satisfies the so called "chained exchangeability property" (see [1], [4]) is center-point complete. This class of spaces includes all uniformly convex spaces and some normed linear spaces of real-valued functions whose norms are in some sense like the supremum-norm (see [4]). Therefore theorem 5 applies, for example, to each of the spaces $C(\Omega)$ (Ω, a compact topological space), c , c_0 and l_∞ with their natural norms.

(4). It was shown in [1] that the space l_1 equipped with several norms (e.g. the natural sum-norm) is center-point complete. Therefore theorem 5 is applicable in these cases, too.

(5). It would be desirable to have some characterizations of normed linear spaces which are center-point complete. A few sufficient and necessary conditions are known (for a survey see [22]) but the general case is still under investigation.

We will now consider certain perturbation problems for generalized contractions.

Lemma 2. Let $(E,\|\ \|)$ be a n.l.s., $X \subset E$ be weakly compact, $u \in X$, $(u_n)_{n\in \mathbb{N}} \in X^{\mathbb{N}}$ and $f\colon X \to E$ be a generalized contraction such that

$$\lim(u_n)_{n\in \mathbb{N}} = u \quad \text{(weakly)} \tag{*}$$

$$\lim(u_n - f(u_n))_{n\in \mathbb{N}} = 0 \quad \text{(strongly)} \tag{**}$$

Let furthermore one of the following conditions be satisfied:

$$f[\partial X] \subset X \tag{2.1}$$

$$f \text{ is weakly sequentially continuous} \tag{2.2}$$

$$(E, \| \ \|) \text{ is uniformly convex and } X \text{ is convex} \tag{2.3}$$

$$\forall_{(x_n)_{n\in \mathbb{N}} \in E^{\mathbb{N}}} \ \forall_{x,y \in E} \ [\lim(x_n)_{n\in\mathbb{N}} = x \ \text{(weakly)} \Rightarrow \Rightarrow \underline{\lim}(\|x_n - x\|)_{n\in\mathbb{N}} \leq \underline{\lim}(\|x_n - y\|)_{n\in\mathbb{N}}] \tag{2.4}$$

(If (2.4) holds, $(E, \| \ \|)$ is said to satisfy a weak OPIAL-condition (see [12], [23])). Then $f(u_n)_{n\in\mathbb{N}} = u$ (strongly) .

Proof (2.1). By lemma 1 (ii) there is $a \in E$ such that $a \in C(X, (u_n)_{n\in\mathbb{N}})$. Lemma 1 (iii) yields by (**): $R((u_n)_{n\in\mathbb{N}})(f(a)) \leq R((u_n)_{n\in\mathbb{N}})(a)\cdot\alpha(a)$. Using lemma 1(i) and $f[\partial X] \subset X$ we see that $R((u_n)_{n\in\mathbb{N}})(a) = 0$. Therefore $\lim(u_n)_{n\in\mathbb{N}} = a$ (strongly) and $f(a) = a$. By (*) we have $u = a$. (2.2), (2.3), (2.4): We first prove $f(u) = u$. In the case of (2.2) this is trivial and in the case of (2.3) this is well-known ($I-f$ is demi-closed). Given (2.4) we have

$$\begin{aligned}
\underline{\lim}(\|u_n - f(u)\|)_{n\in\mathbb{N}} &\leq \underline{\lim}(\|u_n - f(u_n)\| + \|f(u_n) - f(u)\|)_{n\in\mathbb{N}} \\
&\leq \alpha(u) \ \underline{\lim}(\|u_n - u\|)_{n\in\mathbb{N}} \\
&\leq \alpha(u) \ \underline{\lim}(\|u_n - f(u)\|)_{n\in\mathbb{N}} .
\end{aligned}$$

Hence $\underline{\lim}(\|u_n - f(u)\|)_{n\in\mathbb{N}} = 0$ and therefore (by (*)) $f(u) = u$, too. Finally if $f(u) = u$, we obtain for $n \in \mathbb{N}$: $\|u_n - u\| \leq (1-\alpha(u))^{-1}\|u_n - f(u_n)\|$ and using (**) we are done.

Q.E.D.

Remark 5. A reflexive Banach-space which lacks the normal structure property (see [3]) but satisfies the weak OPIAL-condition is the sequence space l_2 renormed by $\|\cdot\| := \max\{\frac{1}{2}\|\cdot\|_2, \|\cdot\|_\infty\}$ (see [13]).

Definition 3. Let $(E, \| \ \|)$ be a n.l.s. and $X \subset E$. X is said to be shrinkable : $\Leftrightarrow$ $[0,1)\ \overline{X} \subset \operatorname{int}(X)$ (see [18]). A convex neighborhood of the origin is a shrinkable set (but the converse is not true). A nonvoid shrinkable set is a starshaped neighborhood of the origin (but the converse is not true).

Theorem 6. Let $(E,\|\ \|)$ be a Banach-space and $\emptyset \neq X \subset E$ be a weakly compact and shrinkable or convex subset of E. Furthermore let $f\colon X \to E$ be a generalized contraction and $g\colon X \to E$ be compact such that

$$\forall_{x,y\in X} \quad [x\in\partial X \Rightarrow f(x)+g(y)\in X] \tag{6.1}$$

Then $\mathrm{Fix}(f+g) \neq \emptyset$.

Proof. Without loss of generality $0 \in X$. By a standard argument (see [19], [24], [30], [31], [32]) there is $(u_n)_{n\in\mathbb{N}} \in X^{\mathbb{N}}$ with $f(u_n)+g(u_n) = (1+\frac{1}{n})u_n$ for $n \in \mathbb{N}$. We may assume the existence of $(u,z) \in X\times E$ with $\lim(u_n)_{n\in\mathbb{N}} = u$ (weakly) and $\lim(g(u_n))_{n\in\mathbb{N}} = z$ (strongly) and conclude $\lim(u_n-f(u_n))_{n\in\mathbb{N}} = z$ (strongly). Defining $h\colon X \to E$ by $h(x) := f(x)+z$ we have $h[\partial X] \subset X$ (according to (6.1)), h is a generalized contraction and $\lim(h(u_n)-u_n)_{n\in\mathbb{N}} = 0$ (strongly). By lemma 2 (hypothesis (2.1) is fulfilled) we have $h(u) = u$ and $\lim(u_n)_{n\in\mathbb{N}} = u$ (strongly). This implies $g(u) = z$ and $f(u)+g(u) = u$.

Q.E.D.

Theorem 7. Let $(E,\|\ \|)$ be a Banach-space and $\emptyset \neq X \subset E$ be weakly compact and convex. Let $f\colon X \to E$ be a generalized contraction and $g\colon X \to E$ be compact such that

$$(f+g)[\partial X] \subset X .$$

Let furthermore one of the following conditions be satisfied:

$$f \text{ is weakly sequentially continuous} \tag{7.1}$$

$$(E,\|\ \|) \text{ is uniformly convex} \tag{7.2}$$

$$(E,\|\ \|) \text{ satisfies the weak OPIAL-condition} \tag{7.3}$$

Then $\mathrm{Fix}(f+g) \neq \emptyset$.

Proof. Without loss of generality $0 \in X$. The main results in the fixed point theory for condensing mappings (see [6], [10], [30], [32]) guarantee the existence of a sequence $(u_n)_{n\in\mathbb{N}} \in X^{\mathbb{N}}$ such that $f(u_n)+g(u_n) = (1+\frac{1}{n})u_n$ for $n \in \mathbb{N}$. By the method given in the proof of theorem 6 and using lemma 2 we get the desired result.

Q.E.D.

Theorem 8. Let $(E,\|\ \|)$ be a Banach-space and $\emptyset \neq X \subset E$ be an open neighborhood of the origin such that $\overline{X}$ is weakly compact.

Let $f: \overline{X} \to E$ be a generalized contraction and $g: \overline{X} \to E$ be compact such that

$$\forall_{x \in \partial X} \ \forall_{\lambda \geq 0} \ [f(x)+g(x) = \lambda x \Rightarrow \lambda \leq 1] \ . \qquad \text{(LS)}$$

Let furthermore one of the following conditions be satisfied:

f is weakly sequentially continuous (8.1)

$(E, \| \ \|)$ is uniformly convex and X is convex (8.2)

$(E, \| \ \|)$ satisfies the weak OPIAL-condition (8.3)

Then $\text{Fix}(f+g) \neq \emptyset$.

Proof. Analogous to that of theorem 7 using a standard result from the fixed point theory for condensing mappings (see [26]).

Q.E.D.

Remark 6 (1). It is not known, whether theorem 7 and theorem 8 are true if the conditions (7.1)-(7.3) and (8.1)-(8.3) are omitted.
(2). Theorems 6, 7 and 8 improve related results due to W.V. Petryshyn [25], [26], J. Reinermann [31], W.A. Kirk [15] and others.

References

[1] Anderson, C.L., W.H. Hyams and C.K. McKnight, Center points of nets, Can. J. Math. 27 (1975) 418-422.

[2] ________, Generalization of a fixed point theorem of M. Edelstein (preprint).

[3] Belluce, L.P., W.A. Kirk and E.F. Steiner, Normal structure in Banach-spaces, Pacific J. Math. 26 (1968) 433-440.

[4] Calder, J.R., W.P. Coleman and R.L. Harris, Centers of infinite bounded sets in a normed space, Can. J. Math. 25 (1973) 986-999.

[5] Caristi, J. and W.A. Kirk, Geometric fixed point theory and inwardness conditions (preprint 1975).

[6] G. Darbo, Punti uniti in transformazioni a condominio non compato, Rend. del Sem. Mat. Univ. Padova 24 (1955) 84-92.

[7] Day, M.M., R.C. James and S. Swaminathan, Normed linear spaces that are uniformly convex in every direction, Can. J. Math. 23 (1971) 1051-1053.

[8] Edelstein, M., The construction of an asymptotic center with a fixed point property, Bull. Amer. Math. Soc. 78 (1972) 206-208.

[9] Edelstein, M., Fixed point theorems in uniformly convex Banach spaces, Proc. Amer. Math. Soc. 44 (1974) 369-380.

[10] Furi, M. and A. Vignoli, On α-nonexpansive mappings and fixed points, Accad. Naz. Lincei 48 (1970) 195-198.

[11] Garkavi, A.L., The best possible net and the best possible cross-section of a set in a normed space, Amer. Math. Soc. Transl. Ser. 2, 39 (1964) 111-132.

[12] Gossez, J.P. and E. Lami Dozo, Some geometric properties related to the fixed point theory for nonexpansive mappings, Pacific J. Math. 40 (1972) 565-573.

[13] Karlowitz, L.A., On nonexpansive mappings, Inst. for Fluid Dynamics and Applied Mathematics, Technical Note BN-805, Univ. of Maryland, College Park (1974).

[14] Kirk, W.A., A fixed point theorem for mappings which do not increase distances, Amer. Math. Monthly 72 (1965) 1004-1006.

[15] Kirk, W.A., On nonlinear mappings of strongly semicontractive type, J. Math. Anal. Appl. 27 (1969) 409-412.

[16] Kirk, W.A., Mappings of generalized contractive type, J. Math. Anal. Appl. 32 (1970) 567-572.

[17] Kirk, W.A., Fixed point theorems for nonexpansive mappings, A.M.S. Nonl. Functional Anal. Vol. 18 (1970) 162-168.

[18] Klee, V., Shrinkable sets in Hausdorff linear spaces, Math. Ann. 141 (1960) 281-285.

[19] Krasnoselskii, M.A., Two remarks on the method of successive approximations, Uspekhi Math. Nauk 10 (1955) 123-127.

[20] Lim, T.C., A fixed point theorem for multivalued nonexpansive mapping in a uniformly convex Banach space, Bull. Amer. Math. Soc. 80 (1974) 1123-1126.

[21] Lim, T.C., Characterization of normal structure, Proc. Amer. Math. Soc. 43 (1974) 313-319.

[22] Müller, G., Cl. Krauthausen, J. Reinermann, R. Schöneberg, New fixed point theorems for compact and nonexpansive mappings and applications to Hammerstein-equations, Sonderforschungsbereich 72 an der Universität Bonn, preprint, to appear.

[23] Opial, Z., Weak convergence of the sequence of successive approximations for nonexpansive mappings, Bull. Amer. Math. Soc. 73 (1967) 591-597.

[24] Petryshyn, W.V., Remarks on condensing and k-set-contractive mappings, J. Math. Anal. Appl. 39 (1972) 717-741.

[25] Petryshyn, W.V., A new fixed point theorem and its applications Bull. Amer. Math. Soc. 78 (1972) 225-229.

[26] Petryshyn, W.V., Fixed point theorems for various classes of 1-set-contractive and 1-ball-contractive mappings in Banach-spaces, Trans. Amer. Math. Soc. 182 (1973) 323-352.

[27] Reich, S., Remarks on fixed points II, Accad. Naz. Linc. Rend. Sc. fis. mat. e. nat. LIII, (1972) 170-174.

[28] Reinermann, J., Über TOEPLITZsche Iterationsverfahren und einige Anwendungen in der konstruktiven Fixpunkttheorie, Studia Math. 32 (1969) 209-227.

[29] Reinermann, J., Über Fixpunkte kontrahierender Abbildungen und schwach konvergente TOEPLITZ-Verfahren, Arch. d. Math. 20 (1969) 59-64.

[30] Reinermann, J., Fixpunktsätze vom Krasnoselskii-Tye, Math. Z 119 (1971) 339-344.

[31] Reinermann, J., Fixed point theorems for generalized contractions, Comment. Math. Univ. Car. 13 (1972) 617-627.

[32] Reinermann, J. and V. Stallbohm, Fixed point theorems for compact and nonexpansive mappings on starshaped domains, Math. Balkanica 4 (1974) 511-516.

[33] Schöneberg, R., Eine Abbildungsgradtheorie für A-Operatoren mit Anwendungen, Diplomarbeit an der Techn. Hochschule Aachen 1975 (unpublished)

Lehrstuhl C für Mathematik
der Techn. Hoschschule Aachen
Templergraben 55
5100 Aachen
Bundesrepublik Deutschland

SOME RESULTS AND PROBLEMS IN THE FIXED POINT THEORY FOR NONEXPANSIVE AND PSEUDOCONTRACTIVE MAPPINGS IN HILBERT-SPACE

J. Reinermann and R. Schöneberg

Introduction

It is the aim of this paper to give some applications of a simple but powerful convergence lemma established by M.G. Crandall and A. Pazy [7] to the fixed point theory of nonexpansive and pseudocontractive mappings in Hilbert-space. The results obtained by this method strengthen some related theorems due to F.E. Browder [2], [4], [5]; J. Gatica [12], [13]; D. Göhde [14]; W.A. Kirk [12], [17]; W.V. Petryshyn [2], [23], [24]; J. Reinermann and V. Stallbohm [29], [30].

All linear spaces occuring in this paper are assumed to be real linear spaces (without loss of generality). If $(E,||\ ||)$ is a normed linear space, $X \subset E$ and $f: X \to E$, $\overline{X}$ denotes the closure of X and ∂X, $Fix(f) \subset E$ are respectively defined by $\partial X: \overline{X} \setminus int(X)$ and $Fix(f) := \{x | x \in X \wedge f(x) = x\}$.

<u>Definition 1</u>. Let $(E,||\ ||)$ be a normed linear space, let $\emptyset \neq X \subset E$ and $f: X \to E$;

(i) f is said to be <u>nonexpansive</u>: $\Leftrightarrow \forall_{x,y \in X} \ ||f(x)-f(y)|| \leq ||x-y||$;

(ii) f is said to be pseudocontractive: =

$$\Leftrightarrow \forall_{x,y \in X} \ \forall_{\lambda \geq 0} \ ||x-y|| \leq ||(1+\lambda)(x-y) - \lambda(f(x)-f(y))||.$$

In a Hilbert-space $(E,(,))$ with the corresponding norm $||\cdot|| := (\cdot,\cdot)^{1/2}$, (ii) is equivalent to

(iii) $\forall_{x,y \in X} \ ||f(x)-f(y)||^2 \leq ||x-y||^2 + ||(I-f)(x)-(I-f)(y)||^2$,

and this is equivalent to

(iv) $\forall_{x,y \in X} \quad (f(x)-f(y),x-y) \leq ||x-y||^2$.

Lemma 1. (M.G. Grandall and A. Pazy [7]). Let $(E,(,))$ be a Hilbert-space, $\{x_n\} \in E^{\mathbb{N}}$, $\{r_n\} \in (0,\infty)^{\mathbb{N}}$ such that

(i) $\{x_n\}$ is bounded, (ii) $\{r_n\}$ is strictly decreasing,

(iii) $\forall_{n,m \in \mathbb{N}} \quad (r_n x_n - r_m x_m, x_n - x_m) \leq 0$.

Then there exists $x \in E$ such that $\lim\{x_n\} = x$ (strongly).

The proof of lemma 1 is induced by the identity

$$\forall_{n,m \in \mathbb{N}} \quad 2(r_n x_n - r_m x_m, x_n - x_m) = (r_n + r_m)||x_n - x_m||^2 + (r_n - r_m)(||x_n||^2 - ||x_m||^2).$$

It would be desirable to have results of lemma 1 type in the context of more general Banach spaces (l_p $(1 < p < \infty)$, Banach spaces with a weakly continuous duality mapping,...).

A trivial application of lemma 1 to the fixed point theory of pseudo-contractive mappings in Hilbert-spaces is

Lemma 2. Let $(E,(,))$ be a Hilbert-space and $\emptyset \neq X \subset E$; let $f: \bar{X} \to E$ be continuous and pseudocontractive; further let $\{\lambda_n\} \in (0,1)^{\mathbb{N}}$ and $\{x_n\} \in \bar{X}^{\mathbb{N}}$ such that

(i) $\{x_n\}$ is bounded, (ii) $\{\lambda_n\}$ is strictly increasing,

(iii) $\lim\{\lambda_n\} = 1$, (iv) $\forall_{n \in \mathbb{N}} \quad \lambda_n f(x_n) = x_n$.

Then $\mathrm{Fix}(f) \neq \emptyset$.

Proof. Defining $\{r_n\} \in (0,\lambda)^{\mathbb{N}}$ by $r_n := \frac{1}{\lambda_n} - 1$ we get - observing (iv) and definition 1, (iv) - for $n,m \in \mathbb{N}$

$$(r_n x_n - r_m x_m, x_n - x_m) = ((\frac{1}{\lambda_n} - 1)x_n - (\frac{1}{\lambda_n} - 1)x_m, x_n - x_m) =$$

$$= (f(x_n) - f(x_m), x_n - x_m) - ||x_n - x_m||^2 \leq 0 .$$

According to lemma 1 and the identity $f(x_n) - x_n = (1-\lambda_n)f(x_n)$ the conclusion follows.

An application of lemma 2 is

Theorem 1. Let $(E,(,))$ be a Hilbert space, let $X \subset E$ be closed and starshaped (i.e. there exists $x_0 \in X$ such that $(1-t)x_0 + tx \in X$

for $t \in [0,1]$ and $x \in X$), let further $f: X \to E$ be nonexpansive such that

(i) $f[\partial X] \subset X$,

(ii) There exists $\hat{x}_1 \in X$ with bounded Picard-sequence $\{f^n(\hat{x}_1)\}$.

Then $\mathrm{Fix}(f) \neq \emptyset$.

<u>Proof</u>. Without loss of generality we may assume 0 being a star-point of X . Choose $\{\lambda_n\} \in (0,1)^{\mathbb{N}}$ strictly increasing with $\lim\{\lambda_n\} = 1$. Then $\lambda_n f$ is a Banach-contraction satisfying $\lambda_n f[X] \subset X$ (by (i)). According to N.A. Assad [1] there is $\{x_n\} \in X^{\mathbb{N}}$ such that $\lambda_n f(x_n) = x_n$ for $n \in \mathbb{N}$. Using Lemma 2 we are done, if $\{x_n\}$ is shown to be bounded. In order to prove this, let $S := \{f^n(x_1) n \in \mathbb{Z}^+\}$ and K be a closed ball about the origin with $S \subset K$. We claim $x_n \in K$ for $n \in \mathbb{N}$. Indeed, if $\varepsilon > 0$, $x \in X \cap E \setminus K$ and $f(x) = (1+\varepsilon)x$, we obtain by a straightforward computation (Hilbert-space!) for every $y \in S$: $||f(x)-f(y)||^2 \geq$ $\geq \mathrm{dist}((1+\varepsilon)x,S)^2 \geq \mathrm{dist}(x,S)^2 + \varepsilon^2||x||^2$. Thus, choosing $y \in S$ with $\mathrm{dist}(x,S)^2 \geq ||x-y||^2 - \frac{\varepsilon^2}{2}||x||^2$ we find $||f(x)-f(y)||^2 \geq ||x-y||^2 + \frac{\varepsilon^2}{2}||x||^2$, a contradiction to the nonexpansiveness of f . Hence $x_n \in K$ for every $n \in \mathbb{N}$.

<u>Remark 1</u>. Theorem 1 strengthens corresponding results of F.E. Browder [2], D. Göhde [14], J. Reinermann and V. Stallbohm [29], [30]. We do not know, whether Theorem 1 is true for merely lipschitzian pseudocontractive mappings.

It should be noted, that condition (i) of Theorem 1 is unnecessary, if X is assumed to be convex (see [38]).

Using Theorem 1 and a recent result due to W.A. Kirk [18] we get the following.

<u>Corollary 1</u>. Let $(E,(,))$ be a Hilbert-space and let $X \subset E$ be closed, bounded and starshaped. Let $f: X \to E$ be lipschitzian and pseudocontractive such that

$$\underset{x_0 \in X}{\exists} \ \underset{x \in \partial X}{\forall} \ ||x_0 - f(x_0)|| < ||x - f(x)|| \qquad (*)$$

Then $\mathrm{Fix}(f) \neq \emptyset$.

<u>Theorem 2</u>. Let $(E,(,))$ be a Hilbert-space and let $\emptyset \neq X \subset E$ be closed, bounded and convex. Let $f: X \to E$ be continuous and

pseudocontractive such that

$$\forall_{x\in\partial X} \lim_{\lambda\to 0^+} \left\{\frac{\operatorname{dist}((1-\lambda)x+\lambda f(x),X)}{\lambda}\right\} = 0 \qquad (**)$$

(i.e. f is "weakly inward", see B. Halpern and G. Bergman [16] J. Caristi and W.A. Kirk [6]). Then $\operatorname{Fix}(f) \neq \emptyset$.

Proof. Let $0 \in X$ and $\{\lambda_n\} \in (0,1)^{\mathbb{N}}$ be strictly increasing with $\lim\{\lambda_n\} = 1$. Then $\lambda_n f$ satisfies $(\lambda_n f(x)-\lambda_n f(y),x-y) \leq \lambda_n ||x-y||^2$ and (**) too (X is convex). According to R.H. Martin [20] and K. Deimling [8] there is $\{x_n\} \in X^{\mathbb{N}}$ such that $\lambda_n f(x_n) = x_n$. Lemma 2 gives the end of the proof.

Remark 3. Theorem 2 improves related results due to K. Deimling [8] and J. Caristi and W.A. Kirk [6].

Theorem 3. Let $(E,(,))$ be a Hilbert-space and let $X \subset E$ be open and bounded with $0 \in X$. Let $f: \bar{X} \to E$ be lipschitzian and pseudocontractive such that

$$\forall_{x\in\partial X} \ \forall_{\lambda\in\mathbb{R}} [f(x)=\lambda x \Longrightarrow \lambda\leq 1] \text{ (Leray-Schauder condition)} \qquad (***)$$

Then $\operatorname{Fix}(f) \neq \emptyset$.

Proof. Let $\{\lambda_n\} \in (0,1)^{\mathbb{N}}$ be strictly increasing with $\lim\{\lambda_n\} = 1$. There is $\{t_n\} \in (0,1)^{\mathbb{N}}$ such that $f_{t_n} : \bar{X} \to E$ defined by $f_{t_n} := (1-t_n)I+t_n\lambda_n f$ is a Banach-contraction (indeed, a straightforward computation gives $||f_{t_n}(x)-f_{t_n}(y)||^2 \leq ((1-t_n)^2 + 2t_n\lambda_n(1-t_n) + t_n^2\lambda_n^2L^2)||x-y||^2$ for $x,y \in \bar{X}$ and $n \in \mathbb{N}$, L being a Lipschitz constant for f; see also F.E. Browder and W.V. Petryshyn [3]) and satisfies (***) too. According to R.D. Nussbaum's fixed point theory for mappings being k-set-contractions ($0 \leq k < 1$) in the sense of the Kuratowski-measure of noncompactness (see R.D. Nussbaum [21], [22] and W.V. Petryshyn and P.M. Fitzpatrick [25] and W.V. Petryshyn [23]) there is $\{x_n\} \in \bar{X}^{\mathbb{N}}$ such that $f_{t_n}(x_n) = x_n$ for $n \in \mathbb{N}$. This implies $\lambda_n f(x_n) = x_n$ for $n \in \mathbb{N}$ and lemma 2 gives the conclusion.

Remark 4. Theorem 3 strengthens corresponding results due to F.E. Browder [5] and J. Gatica and W.A. Kirk [12]; the latter are forced to assume that $(I-f)(\bar{X})$ is closed.

Definition 2. Let $(E, \|\ \|)$ be a normed linear space; $\emptyset \neq X \subset E$ is said to be shrinkable: $\Leftrightarrow [0,1)\bar{X} \subset \operatorname{int}(X)$ (see V. Klee [19]).

A shrinkable set is a starshaped (with respect to 0) neighborhood of the origin (but the converse is not true).
Theorem 3 implies the following variant of theorem 1:

Corollary 2. Let $(E,(,))$ be a Hilbert-space and let $X \subset E$ be an open bounded shrinkable set. Let $f: X \to E$ be lipschitzian and pseudocontractive such that $f[\partial X] \subset \bar{X}$. Then $\operatorname{Fix}(f) \neq \emptyset$.

Proof. Let $x \in \partial X$ and $\lambda \in \mathbb{R}$ such that $f(x) = \lambda x$. If $p: E \to \mathbb{R}^+$ denotes the Minkowski-functional of X we have $p^{-1}[\{1\}] = \partial X$ and $p^{-1}[[0,1]] = \bar{X}$ (see V. Klee [19]), hence $\lambda \leq 1$, i.e. condition (***) of theorem 3 is satisfied.

Remark 5. If $(E,(,))$ is a Hilbert-space, $\emptyset \neq X \subset E$ is open and bounded, if $k \geq 0$ and $f: \bar{X} \to E$ is both a k-set-contraction in the sense of the Kuratowski-measure of noncompactness Δ and a pseudo-contractive mapping, then we obtain by a computation completely analogous to that indicated in the proof of theorem 3 for t, $\lambda \in (0,1)$ and $f_t := (1-t)I + t\lambda f$ the inequality

$$\underset{D \subset X}{\forall} \Delta(f_t[D]) \leq ((1-t)^2 + 2t\lambda(1-t) + \lambda^2 t^2 k^2)\Delta(D),$$

which implies that for sufficiently small $0 < t < 1$, f_t is a α-set-contraction with $0 < \alpha < 1$. Thus we obtain the following:

Theorem 4. Let $(E,(,))$ be a Hilbert-space and let $X \subset E$ be open and bounded with $0 \in X$. Let $f: \bar{X} \to E$ be a k-set-contraction (in the sense of Kuratowski-measure of noncompactness, $k \geq 0$) and pseudocontractive such that

$$\underset{x \in \partial X}{\forall} \ \underset{\lambda \in \mathbb{R}}{\forall} \ [f(x) = \lambda x \Rightarrow \lambda \leq 1] \qquad (^{**}_{**})$$

Then $\operatorname{Fix}(f) \neq \emptyset$.

This result is related to a corresponding one due to W.V. Petryshyn [24].
Finally let us give a fixed point theorem of antipodal type for lipschitzian pseudocontractive mappings in Hilbert-spaces:

Theorem 5. Let $(E,(,))$ be a Hilbert-space and let $X \subset E$ be an open bounded and symmetric neighborhood of the origin. Let $f\colon \bar{X} \to E$ be lipschitzian and pseudocontractive such that

$$\forall_{x \in \partial X} \quad f(-x) = -f(x) \ . \qquad (+)$$

Then $\mathrm{Fix}(f) \neq \emptyset$.

Proof. By means of Definition 1 (iii) and (+) we obtain for $x \in \partial X$ $\|f(x)\|^2 \leq \|x\|^2 + \|x-f(x)\|^2$, which evidently implies condition $(^{**}_{**})$ of Theorem 4.

Remark 6. In the same manner as indicated in remark 5, theorem 5 generalizes to pseudocontractive mappings being merely k-set-contractive $(k \geq 0)$.

Let us bring this paper to a close, discussing a fixed point problem originating from theorem 1.

Theorem P (?). Let $(E,\|\ \|)$ be a uniformly convex Banach space, let $k \in \mathbb{N}$ and $\emptyset \neq X_1,\dots,X_k \subset E$ be weakly compact convex subsets of E . Let $Y := \bigcup_{i=1}^{k} X_i$ and assume Y being contractible in the strong topology. Finally let $f\colon Y \to E$ be nonexpansive such that $f[\partial Y] \subset Y$. Then $\mathrm{Fix}\ (f) \neq \emptyset$.

Let us give some comments on this problem:

Remark 7 (1). Theorem P is obviously true for $k = 1$ (Browder/Göhde/Kirk Theorem).

(2). Theorem P is true for $k = 2$ (indeed, the contractibility of $X_1 \cup X_2$ implies $X_1 \cap X_2 \neq \emptyset$, hence $X_1 \cup X_2$ is starshaped. Without loss of generality let $0 \in X_1 \cap X_2$. Then for $t \in (0,1)$ $g_t := tf\colon X_1 \cup X_2 \to E$ is a Banach-contraction such that $g_t[\partial(X_1 \cup X_2)] \subset X_1 \cup X_2$. The Assad theorem [1] implies $\mathrm{Fix}(g_t) \neq \emptyset$. Thus choosing a sequence $\{t_n\} \in (0,1)^{\mathbb{N}}$ with $\lim\{t_n\} = 1$ we get a sequence $\{x_n\} \in (X_1 \cup X_2)^{\mathbb{N}}$ such that $t_n f(x_n) = x_n$. Then $x_n - f(x_n) \to 0$ (strongly). We may assume $x_n \in X_1$ for all $n \in \mathbb{N}$ and a fixed $i \in \{1,2\}$ and then $x_n \to x \in X_i$ (weakly). The demiclosedness of $I-f|_{X_i}$ implies $x = f(x)$ and thus $\mathrm{Fix}(f) \neq \emptyset$. Clearly the same argument works in the general case of theorem P, if the contractibility assumption is strengthened to $\bigcap_{i=1}^{k} X_i \neq \emptyset$.

(3). Applying a method due to D. Göhde [14] theorem P is true in the case of a Hilbert-space if the X_i's are closed balls (Y may be assumed to be contractible in the strong or weak topology of $(E,(,))$), using the following fixed point theorem of Rothe-type:

Theorem. Let $k \in \mathbb{N}$ and $\emptyset \neq X \subset \mathbb{R}^k$ be a compact, contractible ANR. Let $f: X \to \mathbb{R}^k$ be continuous such that $f[X] \subset X$. Then $Fix(f) \neq \emptyset$.

For a proof and a more general theorem of this type see [38]. This theorem extends the well-known Knaster/Kuratowski/Mazurkiewicz fixed point theorem [34] (which is also true in infinite-dimensional Banach spaces, see J. Reinermann [28]).

(4). Theorem P is true - even without any contractibility condition - if f satisfies $||f(x)-f(y)|| < ||x-y||$ for $x,y \in X$ with $x \neq y$ (see [38]).

(5). In the case of a Hilbert-space $(E,(,))$ theorem P was proved by D. Göhde [14] under the additional assumptions that $f[Y] \subset Y$ and the X_i's are closed balls (without loss of generality f can be assumed to be defined on the whole of E (see B. Grünbaum [15])). For a generalization we need the following result, recently proved by M. Furi and M. Martelli [33]:

Theorem. Let $(E,|| \, ||)$ be a Banach space and let $\emptyset \neq Y \subset E$ be a finite union of bounded, closed and convex subsets of E . Let $f: Y \to Y$ be continuous and a 1-set-contraction with respect to the Kuratowski measure of noncompactness, such that the Lefschetz-number $\Lambda(f) \neq 0$ (defined with respect to the singular homology over the rationals). Then $\inf(\{||x-f(x)|| \mid x \in Y\}) = 0$.
(The proof depends heavily on a corresponding result due to R.D. Nussbaum [36]). Thus we have the following extension of the Browder/Göhde/Kirk theorem, which is also a partial answer to the problem "theorem P".

Theorem 6. Let $(E,|| \, ||)$ be a uniformly convex Banach space, let $k \in \mathbb{N}$ and $X_1,\ldots,X_k$ be nonempty weakly compact convex subsets of E.

Let $Y := \bigcup_{i=1}^{k} X_i$ be contractible in the strong topology and let $f : Y \to Y$ be nonexpansive. Then $\mathrm{Fix}(f) \neq \emptyset$.

Proof. The contractibility of Y implies $\Lambda(f) = 1$. The Furi/Martelli theorem then guarantees the existence of $\{x_n\} \in Y^{\mathbb{N}}$ such that $x_n - f(x_n) \to 0$ (strongly) (f is a 1-set-contraction). We may assume that $x_n \in X_1$ for all $n \in \mathbb{N}$ and a fixed $i \in \{1,\ldots,k\}$ and then $x_n \to x \in X_i$ (weakly). The demiclosedness of $I-f|_{X_i}$ implies $x = f(x)$. Clearly theorem 6 remains true if f is assumed to be the sum of a nonexpansive and a completely continuous operator, or, more generally, if f is assumed to be a LANE-map in the sense of R.D. Nussbaum [35]: Indeed, for LANE-maps, f , we have both, f is a 1-set-contraction and $I-f$ is demiclosed on weakly compact convex sets in a uniformly convex Banach space (see V. Stallbohm [37] and R.D. Nussbaum [35]).

References

[1] Assad, N.A., A fixed point theorem for weakly uniformly strict contractions, Can. Math. Bull. 16 (1973) 15-18.

[2] Browder, F.E. and W.V. Petryshyn, The solution by iteration of nonlinear functional equations in Banach spaces, Bull. Amer. Math. Soc. 72 (1966) 571-575.

[3] Browder, F.E. and W.V. Petryshyn, Construction of fixed points of nonlinear mappings in Hilbert spaces, J. Math. Anal. Appl. 20 (1967) 197-228.

[4] Browder, F.E., Nonlinear mappings of nonexpansive and accretive type, Bull. Amer. Math. Soc. 73 (1967) 875-882.

[5] Browder, F.E., Semicontractive and semiaccretive nonlinear mappings in Banach spaces, Bull. Amer. Math. Soc. 74 (1968) 660-665.

[6] Caristi, J. and W.A. Kirk, Geometric fixed point theory and inwardness conditions (preprint (1975)).

[7] Crandall, M.G. and A. Pazy, Semi-groups of nonlinear contractions and dissipative sets, J. Functional Anal. 3 (1969) 376-418.

[8] Deimling, K., Zeros of accretive operators, Manuscripta Math. 13 (1974) 365-374.

[9] Edelstein, M., On fixed and periodic points under contractive mappings, J. Lond. Math. Soc. 37 (1962) 74-79.

[10] Edelstein, M., On nonexpansive mappings in Banach spaces, Proc. Cambridge Phil. Soc. 60 (1964) 439-447.

[11] Edelstein, M. and A.C. Thompson, Contractions, isometries and some properties of inner-product spaces, Indag. Math. 29 (1967) 326-331.

[12] Gatica, J. and W.A. Kirk, Fixed point theorems for lipschitzian pseudocontractive mappings, Proc. Amer. Math. Soc. 36 (1972) 111-115.

[13] Gatica, J., Fixed point theorems for k-set-contractions and pseudocontractive mappings, J. Math. Anal. Appl. 46 (1974) 555-564.

[14] Göhde, D., Elementare Bemerkungen zu nichtexpansiven Selbstabbildungen nicht konvexer Mengen im Hilbertraum, Math. Nachr. 63 (1974) 331-335.

[15] Grünbaum, B., On a theorem of Kirszbraun, Bull. Res. Council Israel 7 (1958) 129-132.

[16] Halpern, B. and G. Bergman, A Fixed point theorem for inward and outward maps, Trans. Amer. Math. Soc. 130 (1968) 353-358.

[17] Kirk, W.A., Remarks on pseudocontractive mappings, Proc. Amer. Math. Soc. 25 (1970) 820-823.

[18] Kirk, W.A., Some results on pseudocontractive mappings (preprint 1975).

[19] Klee, V., Shrinkable neighborhoods in Hausdorff linear spaces, Math. Ann. 141 (1960) 281-285.

[20] Martin, R.H., Differential equations on closed subsets of a Banach space, Trans. Amer. Math. Soc. 179 (1973) 399-414.

[21] Nussbaum, R.D., The fixed point index and fixed point theorems for k-set-contractions, Ph.D. Thesis, Univ. of Chicago, Chicago, Ill. (1969).

[22] Nussbaum, R.D., Degree theory for local condensing maps, J. Math. Anal. Appl. 37 (1972) 741-766.

[23] Petryshyn, W.V., Remarks on condensing and k-set-contractive mappings, J. Math. Anal. Appl. 39 (1972) 717-741.

[24] Petryshyn, W.V., Fixed point theorems for various classes of 1-set-contractive and 1-ball-contractive mappings in Banach spaces, Trans. Amer. Math. Soc. 182 (1973) 323-352.

[25] Petryshyn, W.V. and P.M. Fitzpatrick, Degree theory for noncompact multivalued vector fields, Bull. Amer. Math. Soc. 79 (1973) 609-613.

[26] Reich, S., Fixed points in locally convex spaces, Math. Z. 125 (1972) 17-31.

[27] Reinermann, J., Fixpunktsätze von Krasnoselski-Typ, Math. Z. 119 (1971) 339-344.

[28] Reinermann, J., Neue Existenz- und Konvergenzsätze in der Fixpunkttheorie nichtlinearer Operatoren, J. Approximation Theory (1973) 387-399.

[29] Reinermann, J., and V. Stallbohm, Fixed point theorems for compact and nonexpansive mappings on starshaped domains, Comment. Math. Univ. Carolinae 4 (1974) 775-779.

[30] Reinermann, J. and V. Stallbohm, Fixed point theorems for compact and nonexpansive mappings on starshaped domains, Math. Balkanica 4 (1974) 511-516.

[31] Reinermann, J., Fixed point theorems for nonexpansive mappings on starshaped domains, Berichte der Gesellschaft f. Math. und Datenverarbeitung, Bonn, Nr. 103 (1975) 23-28.

[32] Browder, F.E., Fixed point theorems on infinite-dimensional manifolds, Trans. Amer. Math. Soc. 119 (1965) 179-194.

[33] Furi, M., and M. Martelli, On the minimal displacement under acyclic-valued maps defined on a class of ANR's, Sonderforschungsbereich 72 an der Universität Bonn, preprint No. 39 (1974).

[34] Knaster, B., K. Kuratowski and S. Mazurkiewicz, Ein Beweis des Fixpunktsatzes für n-dimensionale Simplexe, Fund. Math. 14 (1929) 132-137.

[35] Nussbaum, R.D., The fixed point index and fixed point theorems for k-set-contractions, Ph.D. Thesis, Univ. of Chicago (1969).

[36] Nussbaum, R.D., The fixed point index for local condensing maps, Ann. Mat. Pura Appl. 89 (1971) 217-258.

[37] Stallbohm, V., Fixpunkte nichtexpansiver Abbildungen, Fixpunkte kondensierender Abbildungen, Fredholm'sche Sätze linearer kondensierender Abbildungen, Dissertation an der Technischen Hochschule Aachen (1973).

[38] Müller, G., Cl. Krauthausen, J. Reinermann, R. Schöenberg, New fixed point theorems for compact and nonexpansive mappings and applications to Hammerstein-equations, Sonderforschungsbereich 72 an der Universität Bonn, preprint, to appear.

MAPS OF CONTRACTIVE TYPE[1]

Chi Song Wong

On basic result of maps of contractive type is the Banach contractive mapping theorem: If f is a self map on a (non-empty) complete metric space (X,d) for which there exists k in $[0,1)$ such that $d(f(x), f(y)) \leq kd(x,y)$ for all x,y in X, then (a) f has a unique fixed point x_0 in X; (b) $\{f^n(x)\}$ converges uniformly to x_0 for any bounded subset D of X. The above theorem has pretty statement, easy proofs, important applications and quite a few extensions. We shall limit ourselves to some of these extensions to which we have contributions.

In 1962, M. Edelstein [5] obtained a result which contains the following result as a special case.

Theorem A. Let f be a self map on a compact metric space (X,d). Suppose that f is contractive $(d(f(x), f(y)) < d(x,y)$ for all distinct x,y in $X)$. Then f has a unique fixed point x_0 and $\{f^n(x)\}$ converges to x_0 for all x in X.

The existence of x_0 in Theorem A can be easily seen to follow from the fact that the function V: $V(x) = d(x,f(x))$, $x \in X$, is lower semicontinuous and $V(f(x)) < V(x)$ whenever $V(x) > 0$. This can be seen from the following generalization of the above result and Theorem 1 in [15].

Proposition 1. Let f be a self map on a countably compact topological space X. Suppose that there exists a lower semi-continuous function V of X into $[0,\infty)$ such that whenever $x \neq f(x)$, $V(y) < V(x)$ for some y in X. Then f has a fixed point.

Theorem A is restrictive in the sense that it requires that X is compact. In 1962, E. Rakotch [7] obtained the following result.

Theorem B. Let f be a self map on a complete metric space (X,d). Suppose that there exists an increasing function $0 < \alpha < 1$ on $[0,\infty)$ such that

$$d(f(x),\ f(y)) \leq \alpha(d(x,y))d(x,y)\ ,\ x,y \in X\ .$$

Then f has a unique fixed point x_0 and $\{f^n(x)\}$ converges to x_0 for all x in X .

Obviously, Theorem B generalizes part (a) of the Banach contraction mapping theorem. In 1969, this result was further generalized by D.W. Boyd and J.S.W. Wong [2].

Theorem C. Let f be a self map on a complete metric space (X,d) . Suppose that there exists a self map Φ on $[0,\infty)$ such that Φ is upper semicontinuous from the right, $\Phi(t) < t$ for $t > 0$ and f is Φ-contractive:

$$d(f(x),f(y)) \leq \Phi(d(x,y))\ ,\ x,y \in X\ .$$

Then f has a unique fixed point x_0 and $\{f^n(x)\}$ converges to x_0 for all x in X .

When X is metrically convex, the upper semicontinuity of Φ in Theorem C is not needed and one may assume that Φ is subadditive and use our theorem in [10] to prove that Theorem B and C are equivalent. The proof was simplified by J.S.W. Wong in 1974 [16].

In 1969, E. Keeler and A. Meir [6] obtained the following generalization of Theorem C.

Theorem D. Let f be a self map on a complete metric space (X,d) . Suppose that for any $\varepsilon > 0$, there exists $\delta(\varepsilon) > 0$ such that

$$d(f(x),f(y)) < \varepsilon \quad \text{whenever} \quad \varepsilon \leq d(x,y) < \varepsilon+\delta(\varepsilon)\ .$$

Then f has a unique fixed point x_0 in X and $\{f^n(x)\}$ converges to x_0 for all x in X .

We [14] shall show that Theorem D can be restated so that one can estimate the errors in the successive approximations of the fixed point in terms of a function w which we shall refer to as a control function.

Lemma 1. Let g be an increasing extended real-valued nonnegative function on $[0,\infty)$ such that $g(\varepsilon) > \varepsilon$ for all $\varepsilon \in (0,\infty)$. Then there exists an increasing self map h on $[0,\infty)$ such that $h(0) = 0$, $h \leq g$, h has countable range, $h(\varepsilon) > \varepsilon$ for all $\varepsilon > 0$ and h is is continuous from the right.

Indications of a proof. Let

$$j(x) = \min\{g(x), 2x\} \ , \ x \in [0,\infty) \ .$$

Then j has the given properties of g. So we may assume further that g is real-valued and $\lim_{x \to 0^+} g(x) = 0$. Let $c \in (0,\infty)$ and define $c_0 = c$. Then there exists a unique countable limiting ordinal γ and a unique net $\{c_\beta\}$ of $[0,\gamma)$ into $[0,\infty)$ such that

(a) $c_{\alpha+1} = g(c_\alpha)$ for all $\alpha \in [0,\gamma)$,

(b) $\sup_{\beta<\alpha} c_\beta = c_\alpha$ for every limiting ordinal α in $[0,\gamma)$,

(c) $\sup_{\beta<\gamma} c_\beta = \infty$.

Now let

$$g_c(x) = g(x) \quad \text{for} \quad 0 \le x < c \ .$$

$$g_c(x) = c_{\beta+1} \quad \text{for} \quad c_\beta \le x < c_{\beta+1} \ , \ \beta \in [0,\gamma) \ .$$

Then g_c is an increasing self map on $[0,\infty)$ such that $g_c \le g$, $g_c(\varepsilon) > \varepsilon$ for all $\varepsilon > 0$ and on $[c,\infty)$, g_c has countable range and is continuous from the right. Let

$$h(0) = 0 \ ,$$

$$h(s) = g_{\frac{1}{n+1}}(s) \quad \text{for} \quad \frac{1}{n+1} \le s < \frac{1}{n} \ , \ n = 1,2,\ldots \ ,$$

$$h(s) = g_1(s) \quad \text{for} \quad s \ge 1 \ .$$

Then h is a required function. q.e.d.

Theorem 1. Let f be a self map on a complete metric space (X,d) . Then the following conditions are equivalent:

(i) For any $\varepsilon > 0$, there exists $\delta(\varepsilon) > 0$ such that $d(f(x),f(y)) < \varepsilon$ whenever $\varepsilon \le d(x,y) < \varepsilon + \delta(\varepsilon)$ ....(*)

(ii) There exists an extended real-valued increasing non-negative function w on $[0,\infty)$ such that $w(s) > s$ for $s > 0$ and $w(d(f(x),f(y))) \le d(x,y)$, $x,y \in X$.

(iii) There exists an increasing self map w on $[0,\infty)$ such that w is continuous from the right, $w(s) > s$ for $s > 0$ and $w(d(f(x),f(y))) \le d(x,y)$, $x,y \in X$.

(iv) There exists a self map w on $[0,\infty)$ such that for $s > 0$, $w(s) > s$ and w is lower semicontinuous from the right and

$w(d(f(x),f(y))) \le d(x,y)$, $x, y \in X$. Moreover, if (iv) is satisfied then

(a) f has a unique fixed point,

(b) $\{f^n(x)\}$ converges to x_0 for all x in X .

Indication of a proof. (i) $\Rightarrow$ (ii) . Let $\varepsilon > 0$, $w(0) = 0$, $w(\varepsilon) = \sup\{\varepsilon+\delta(\varepsilon) : \delta(\varepsilon)$ satisfies $(*)\}$. Then w is a required function in (ii). "(ii) $\Rightarrow$ (iii)" follows from Lemma 1. "(iii) $\Rightarrow$ (iv) $\Rightarrow$ (i)" is straightforward. Since (i) $\Rightarrow$ (iv), by Theorem D, (iv) implies (a) and (b). For this fixed point theorem, we can also give a proof that is similar to the proof of Theorem 1 in [2] q.e.d. (We can also prove (ii) $\Rightarrow$ (iii) without using transfinite induction).

In Theorem 1, let A be the set of all points s in $[0,\infty)$ for which there exists a decreasing sequence $\{s_n\}$ in $\{d(f^n(x),f^m(x)) : n,m \ge 0$, $x \in X\}$ that converges to s . Theorem 1 remains true when w in (iv) is merely defined on A , because w can be extended to $[0,\infty)$ without losing its properties needed in (iv). Similar remarks apply to Theorem C.

Intuitively, the difference between Theorem C and the fixed point theorem obtained from (iv) in Theorem 1 is a matter of imposing the control functions on the left hand side or right hand side of certain inequalities. However, there is no symmetry as "left and right" in the sense that Theorem C is a special case of, and is not equivalent to the fixed point theorem obtained from (iv) in Theorem 1[6]. Nevertheless, in a more restrictive situation, there does exist a natural correspondence between Φ in Theorem C and w in Theorem 1. To see this, we need a number of lemmas, each of which, like Lemma 1, is interesting in its own right.

Lemma 2. Let g be a self map on $[0,\infty)$ such that $g(0) = 0$, $g(s) > s$ for all $s > 0$. Let

$$h(t) = \sup\{s \ge 0: g(s) \le t\} , \quad t \in [0,\infty) .$$

Then h is an increasing self map on $[0,\infty)$, $h(0) = 0$ and

(a) $h(g(s)) \ge s$ for all $s \ge 0$;

(b) $g(h(t)) \le t$ for all $t \ge 0$ if g is lower semi-continuous from the left;

(c) $h(t) < t$ for $t > 0$ if g is lower semicontinuous from the left;

(d) h is continuous from the right if g is lower semicontinuous (on $0,\infty)$;

(e) h is continuous from the left if g is not constant on any nonempty open intervals;

(f) $g(h(t)) = t$ for all $t > 0$ if g is lower semicontinuous from the left and upper semicontinuous from the right on $(0,\infty)$.

Lemma 3. Let h be a self map on $[0,\infty)$ such that $h(0) = 0$, $h(t) < t$ for $t > 0$. Let

$$g(s) = \inf\{t \geq 0 : h(t) \geq s\} \ , \ s \in [0,\infty) \ .$$

Then g is an increasing function of $[0,\infty)$ into $[0,\infty]$, $g(0) = 0$ and

(a) $g(h(t)) \leq t$ for all t in $[0,\infty)$;

(b) $h(g(s)) \geq s$ for all s with $g(s) < \infty$ if h is upper semicontinuous from the right on $(0,\infty)$;

(c) $g(s) > s$ for all s in $(0,\infty)$ if h is upper semicontinuous from the right on $(0,\infty)$;

(d) g is continuous from the left if h is upper semicontinuous on $(0,\infty)$;

(e) g is continuous from the right if h is not constant on any nonempty open intervals;

(f) $h(g(s)) = s$ for all s in $[0,\infty)$ if on $(0,\infty)$, h is upper semicontinuous from the right and lower semicontinuous from the left.

Lemma 4. Let h be a self map on $[0,\infty)$ such that $h(0) = 0$, $h(t) < t$ for $t > 0$ and h is upper semicontinuous on $(0,\infty)$. Then there exists a self map ψ on $[0,\infty)$ such that $\psi(0) = 0$ $\psi(t) < t$ for $t > 0$, ψ is increasing, continuous and $\psi > h$.

Indications of a proof. Let $h_1(t) = \max\{h(t), \frac{1}{2}t\}$, $t \geq 0$. Then h_1 has the given properties of h . So we may assume that $\lim_{t\to\infty} h(t) = \infty$. Let g be defined as in Lemma 3. Then g is real-valued. By Lemma 3, $g(0) = 0$, g is increasing, continuous from the left and $g(s) > s$ for $s > 0$. Let h_2 be the function h in

Lemma 2. Then h_2 has the given properties of h, $h \leq h_2$ and h_2 is continuous from the right. So we may assume further that h is increasing and continuous from the right. Let $0 < a < b < \infty$,

$$L(a,b) = \inf_{t\in[a,b]}(t-h(t)) \quad .$$

Then by the lower semicontinuity of $t \to t - h(t)$ on $[a,b]$, $L(a,b) > 0$. Also for any $a > 0$, $L(a, \ )$ is decreasing on (a,∞); for any $b > 0$, $L(\ ,b)$ is increasing and nonexpansive on $(0,b)$. Now let

$$\psi(0) = 0 ;$$

$$\psi(t) = t - L(\frac{1}{n+2},1) , \frac{1}{n+2} < t < \frac{1}{n+1} - (L(\frac{1}{n+1},1) - L(\frac{1}{n+2},1)),$$

$$\psi(t) = \frac{1}{n+1} - L(\frac{1}{n+1},1) , \frac{1}{n+1} - (L(\frac{1}{n+1},1) - L(\frac{1}{n+2},1)) \leq t < \frac{1}{n+1} ,$$

$$n = 1,2,\ldots,$$

$$\psi(t) = t - L(\frac{1}{2},2) , \frac{1}{2} \leq t < 1 ;$$

$$\psi(t) = 2n - L(\frac{1}{2},2n+2) + (1+L(\frac{1}{2},2n) - L(\frac{1}{2},2n+2))(t-2n)$$

$$\text{for } 2n-1 \leq t < 2n ,$$

$$\psi(t) = t - L(\frac{1}{2},n+2) \quad \text{for } 2n \leq t < 2n+1 ,$$

$$n = 1, 2, \ldots .$$

Then ψ is a required function. q.e.d.

Lemma 5. Let g be a self map on $[0,\infty)$ such that $g(0) = 0$, $g(s) > s$ for $s > 0$ and g is lower semicontinuous on $(0,\infty)$. Then there exists a self map ψ on $[0,\infty)$ such that $\psi(0) = 0$. $\psi(s) > s$ for $s > 0$, ψ is increasing, continuous and $\psi \leq g$.

Theorem 2. Let f be a self map on a complete metric space (X,d). Then the following conditions are equivalent:

(i) There exists a self map Φ on $[0,\infty)$ such that $\Phi(t) < t$ for $t > 0$, Φ is increasing, continuous and f is Φ-contractive.

(ii) There exists a self map Φ on $[0,\infty)$ such that $\Phi(t) < t$ for $t > 0$, Φ is upper semicontinuous on $(0,\infty)$ and f is Φ-contractive.

(iii) There exists a self map w on $[0,\infty)$ such that $w(s) > s$ for $s > 0$, w is increasing, continuous and

$$w(d(f(x),f(y))) \leq d(x,y) , \quad x,y\in X .$$

(iv) There exists a self map w on $[0,\infty)$ such that $w(s) > s$ for $s > 0$, w is lower semicontinuous and

$$w(d(f(x),f(y))) \leq d(x,y) \ , \ x,y \in X \ .$$

Hence if any one of (i), (ii), (iii) or (iv) is satisfied, then

(a) f has a unique fixed point x_0,

(b) $\{f^n(x)\}_{x \in D}$ converges uniformly to x_0 for any bounded subset D of X.

<u>Indications of a proof</u>. "(ii) ⇒ (i)" follows from Lemma 4. "(i) ⇒ (ii)" is obvious. "(iv) ⇒ (iii) follows from Lemma 5. "(iii) ⇒ (iv)" is obvious. "(i) ⇒ (iv)" follows from Lemma 3. "(iii) ⇒ (ii)" follows from Lemma 2. "(i) ⇒ (a) and (b)" can be proved in a similar way as Theorem 1 in [3]. q.e.d.

Applying Lemmas 2 and 3, we obtain conditions under which Theorem 2 is applicable.

<u>Theorem 3</u>. Let f be a self map on a complete metric space (X,d). Then the following two conditions are equivalent:

(i) There exists a self map Φ on $[0,\infty)$ such that $\inf_{t \geq a}(t-\Phi(t))$ is positive for all $a > 0$ and t is Φ-contractive.

(ii) There exists a self map w on $[0,\infty)$ such that $\inf_{s \geq b}(w(s)-s)$ is positive for all $b > 0$ and

$$w(d(f(x),f(y))) \leq d(x,y) \ , \ x,y \in X \ .$$

Moreover, each of (i) and (ii) implies any of (i) - (iv) in Theorem 2 and hence implies

(a) f has a unique fixed point x_0;

(b) $\{f^n(x)\}_{x \in D}$ converges uniformly to x_0 for any bounded subset D of X.

The fixed point theorem obtained above generalizes the Banach contraction mapping theorem and if X is bounded, is equivalent to the fixed point theorem obtained in Theorem 2. Details of Theorems 1 - 2 can be found in [14].

All theorems presented above deal with maps whose fixed points are unique and can be approximated by iterates. If we demand merely the uniqueness of fixed point for a given map, many theorems above can be generalized further in a meaningful way. The following result was obtained by the author in 1969 with some help from M.M. Day [9].

Theorem 4. Let f be a self map on a complete metric space (X,d). Suppose that

(a) the function V defined by

$$V(x) = d(x,f(x)) \ , \ x \in X \ ,$$

is lower semicontinuous;

(b) $\inf\{d(x,f(x)) : x \in X\} = 0$;

(c) f is uniformly continuous from $(x, {}_f d)$ into (X,d), where ${}_f d$ is the pseudo metric for X determined by

$${}_f d(x,y) = \max\{d(x,f(x)), d(y,f(y))\} \text{ for } x,y \in X \ , \ x \neq y \ .$$

Then f has a unique fixed point.

By Theorem 3 above and Lemma 2 in [11], we know that Theorem 4 generalizes and hence gives a new proof for part (a) of Theorem 3. Theorem 4 generalizes quite a few results obtained earlier [1], [2]. It was further generalized to the case where X is a complete Hausdorff uniform space [12]. In 1973, Dugundji [4] generalized condition (c) in Theorem 4 as follows: Let (X,d) be a metric space. Let $A \subset X$. Let V be a function of X into $[0,\infty)$. V is said to be positive definite mod A if for all $\varepsilon > 0$,

$$\inf\{V(x) : x \in X \ , \ d(x,A) \geq \varepsilon\} > 0 \ .$$

Theorem E. Let (X,d) be a metric space. Then (X,d) is complete if and only if for any lower semicontinuous function V of X into $[0,\infty)$, $V(x) = 0$ for some x in X whenever $\inf V(x) = 0$ and $(x,y) \to V(x) + V(y)$ is positive definite mod the diagonal of $X \times X$.

The relation between the above result of Dugundji and Theorem 4 was discussed in [13]. In 1974, we [13] generalized Dugundji's notion of positive definiteness as follows. Let $(X,\mathcal{U})$ be a uniform space. Let $A \subset X$. Let X be a function of X into $[0,\infty)$. V is said to be positive definite mod A if for any U in $\mathcal{U}$, $\inf V(X \backslash U[A]) > 0$. K.K. Tan and the author [8], [13] obtained the following generalization of Theorem E.

Theorem 5. Let $(X,\mathcal{U})$ be a uniform space. Then $(X,\mathcal{U})$ is sequentially complete if and only if for any lower semicontinuous function V of X into $[0,\infty)$, $V(x) = 0$ for some x in X whenever $\inf V(X) = 0$

and $(x,y) \to V(x) + V(y)$ is positive definite mod the diagonal of $X \times X$.

Because of a lack of the notion of sequential completions, one must prove the completeness in Theorem 5 by a different method used by Dugundji for the metric spaces setting.

The following generalization of Theorem 4 follows from Theorem 5.

Theorem 6. Let f be a self map on a sequentially complete Hausdorff uniform space induced by a family $\mathcal{D}$ of pseudo metrices for X. For each d in $\mathcal{D}$, let V_d be the map on X such that $V_d(x) = d(x,f(x))$, $x \in X$. Suppose that

(a) For any d in $\mathcal{D}$, V_d is lower semicontinuous on X;

(b) there exists a sequence $\{x_n\}$ in X such that $\{V_d(x_n)\}_{d \in \mathcal{D}}$ converges uniformly to 0;

(c) for any $d \in \mathcal{D}$, $\varepsilon > 0$, there exists $\delta(\varepsilon) > 0$ such that $d(x,y) < \varepsilon$ whenever $V_d(x) + V_d(y) < \delta(\varepsilon)$.

Then f has a unique fixed point.

Dugundji also succeeded in using his notion of positive definiteness to characterize compact subsets of a metric space.

Theorem F. Let A be a nonempty closed subset of a metric space (X,d). The following conditions are equivalent:

(a) A is compact.

(b) For any lower semicontinuous function V of X into $[0,\infty)$, $V(x) = 0$ for some x in A whenever $\inf V(X) = 0$ and V is positive definite mod A.

When (X,d) in Theorem F is replaced by a uniform space $(X,\mathcal{U})$, Tan and the author [8], [13] proved that (i) (a) implies (b) and (ii) (b) implies that A is countably compact; it is however unknown to us if (a) and (b) remain equivalent. In view of the following theorem [8] which gives more general result than (ii), one might suspect the naturality of the role of positive definiteness in Theorem F.

Theorem 7. Let A be a closed subset of a topological space $(X,\mathcal{T})$. Then A is countably compact if and only if for any lower semicontinuous function V of X into $[0,\infty)$ $\inf V(A) = 0$ implies that $V(x) = 0$ for some x in A.

The above result can be proved directly or by its special case:

$A = X$, which can essentially be found in the appendix of [1]. As a simple application, we note that Proposition 1 can be proved by Theorem 7 with $A = X$. We conclude our discussion on an extension of Proposition 1 for the case when X is a complete metric space. We emphasize here that in Propositions 1 and 2, it is not required that f is continuous.

Proposition 2. Let f be a self map on a complete metric space (X,d). Suppose that there exists a lower semicontinuous function of X into $[0,\infty)$ such that for any x in X with $x \neq f(x)$, there exists y in $X - \{x\}$ such that

$$d(x,y) \leq V(x) - V(y) .$$

Then f has a fixed point.

The above result follows easily form the following result of Dr. Caristi, which was communicated to us by Professor W.A. Kirk (Trans. Amer. Math. Soc. 215(1976), 241-251). We can give a shorter proof (Proc. Amer. Math. Soc. (1976)) by refining the transfinite argument given by Caristi. We note also that Theorem G follows from Proposition 2 and a generalization of part (a) of the Banach contraction mapping theorem.

Theorem G. Let f be a self map on a complete metric space (X,d) . Suppose that there exists a lower semicontinuous function V of X into $[0,\infty)$ such that

$$d(x,f(x)) \leq V(x) - V(f(x)), \; x \in X .$$

Then f has a fixed point.

References

[1] Blatter, Jörg, Grothendieck space in approximation theory, Mem. Amer. Math. Soc., No. 120 (1972).

[2] Boyd, D.W., and Wong, J.S.W., On nonlinear contractions, Proc. Amer. Math. Soc., 20(1969), 458-464.

[3] Browder, F.W., On the convergence of successive approximations for nonlinear functional equations, Nederl Akad. Wetensch Proc., Ser A71 = Indag. Math. 30(1968), 27-35.

[4] Dugundji, J., Positive definite functions and coincidence, Séminaire de Mathématiques Supérieures (fixed point theory and its applications) (June 1973).

[5] Edelstein, M., On fixed and periodic points under contractive

mappings, J. London Math. Soc., 37(1962), 74-79.

[6] Keeler, E., and Meir, A., A theorem on contraction mappings, J. Math. Anal. & Appl., 28(1969), 326-329.

[7] Rakotch, E., A note on contraction mappings, Proc. Amer. Math. Soc. 13(1962), 459-465.

[8] Tan, K.K. and Chi Song Wong, On some topological problems arised from mappings of contractive type (submitted).

[9] Wong, Chi Song, Fixed point theorems for nonexpansive mappings, Ph.D. thesis, University of Illinois, Urbana, Illinois, 1969.

[10] __________, Subadditive functions, Pacific J. Math. 36(1971), 549-551.

[11] __________, Fixed point theorems for nonexpansive mappings, J. Math. Anal. & Appl., 37(1972), 142-150.

[12] __________, A fixed point theorem for a class of mappings, Math. Ann., 204 (1973), 97-103.

[13] __________, On Dugundji's notion of positive definiteness, Proc. Amer. Math. Soc. 46 (1974), 443-450.

[14] __________, Characterizations of certain maps of contractive type (submitted).

[15] Wong, J.S.W., Mappings of contractive type on abstract spaces Math. Anal. & Appl., 37 (1972), 331-340.

[16] Wong, J.S.W., A note on subadditive functions, Proc. Amer. Math. Soc. 44(1974), 106.

[1]This research was partially supported by National Research Council of Canada, Grant A8418 and a grant from Canadian Mathematical Congress. This paper is an expanded version of the invited talk I gave at the Seminar on fixed point theory and its applications, Dalhousie University, June 9 - 12, 1975. I would like to thank Professor J. Reinermann for his calling my attention to a paper of Keeler and Meir. With the encouragement of Professor M. Edelstein, I include here some of the results I obtained after the talk.

Department of Mathematics
University of Windsor
Windsor, Ontario, N9B 3P4
Canada

REPORT ON PROBLEMS SESSION

The session was chaired by Les Karlovitz.

Many participants referred to the problems mentioned during their talks. These appear in their respective papers in this volume; in particular, the paper of Edelstein contains a large number of open questions.

In his remarks the chairman referred to the study of a system of nonlinear equations which model the transverse vibrations of a string with masses attached at a finite number of points and with both ends fixed. Referring to R.J. Duffin's work [5] he pointed out how Brouwer's fixed point theorem is used in an unusual and interesting way to show the existence of a periodic solution of the system satisfying the property that, in the space of initial configurations for the masses, the norm of the initial configuration has any prescribed value.

The following additional problems were discussed:

F.E. Browder led the discussion on problems 1, 2 and 3.

1. Classical problem of nonexpansive mappings. Let X be a Banach space, C a convex weakly compact subset of X, and $f: C \to C$ a nonexpansive mapping. Does f have a fixed point?

The papers of M. Edelstein and of Les Karlovitz in this volume are related to this problem.

2. Simultaneous generalization of monotone and accretive mappings. Given Banach spaces X and Y, a continuous function $f: X \to Y$, and a "gauge" function $\phi: X \to Y^*$ (i.e. $\phi(\lambda x) = \lambda\phi(x)$ for $\lambda \geq 0$ and $||\phi(x)||$ is an increasing function of $||x||$) satisfying reasonable conditions such as $||\phi(x)||_{Y^*} = ||x||_X$ and $\phi(x) = Y^*$. One seeks conditions under which strong ϕ-<u>accretiveness</u>, i.e., $\langle f(u)-f(v), \phi(u-v)\rangle \geq c||u-v||^2$ for some $c > 0$ and all $u,v \in X$, implies $f(X) = Y$. In particular, one seeks results which contain the known positive results for <u>monotone mappings</u> (i.e., X reflexive, $Y = X^*$, and ϕ = identity) and for <u>accretive mappings</u>

(i.e., $Y = X$ and ϕ = a duality mapping) and thus provide a link between the two methodologically diverse areas. A discussion of ϕ-accretiveness, with results that require more than the continuity of f are found in Browder [2]. The paper of W. Kirk in this volume is related to some questions of this type.

3. Asymptotic fixed points. Let X be a Banach space and $f: X \to X$ a continuous mapping. Suppose that for some $n > 1$ $f^n(X)$ is relatively compact. Does f have a fixed point? Present results make additional local assumptions on f. For example, (Browder [3]) if f is assumed to be locally compact (i.e., for each point x there is a neighborhood U of x so that $f(U)$ is relatively compact), then f does have a fixed point. For another example, (Nussbaum [13]) it is known that if there is an open neighborhood V of the closure of $f^n(X)$ so that f^n restricted to V is a k-set contraction, $k < 1$, and f restricted to V is continuously Fréchet differentiable, then f has a fixed point.

J. Reinermann led the discussion concerning the following problems:

Notation: $\bar{X}$, ∂X denote the closure and the boundary of a set $X \subset E$ respectively.

4. Let $\emptyset \neq X \subset \mathbb{R}^2$ be compact and starshaped and let $f: X \to \mathbb{R}^2$ be continuous with $f[\partial X] \subset X$. Does f have a fixed point? If $f[X] \subset X$ the answer is yes, see [20], [24]. For $n \geq 3$ the answer is no, see [20], the usual counterexamples are boundary sets in $\mathbb{R}^n$. This leads to the next problem:

5. Let $n \in \mathbb{N}$ and let $\emptyset \neq X \subset R^n$ be open bounded and starshaped. Let $f: \bar{X} \to \mathbb{R}^n$ be continuous such that $f[\partial X] \subset \bar{X}$. Does f have a fixed point? If either $f[X] \subset \text{int}(X)$ or X is shrinkable (see [11]) the answer is yes [20].

6. Let $(E, \|\ \|)$ be an infinite-dimensional Banach space and let S be its unit sphere. Let $f: S^\infty \to S^\infty$ be condensing in sense of the Kuratowski-measure of non-compactness (see [4], [7], [14], [19]). Does f have a fixed point? For compact and, more general, for k-set-contractions ($0 \leq k < 1$) the answer is affirmative, see [13] and the same can be said if f maps into a "semi-sphere" [18]. Obviously

f has a fixed point if the following problem would be positively answered:

7. Let $(E, \|\ \|)$ be an infinite-dimensional Banach space. Let $B^{\infty} := \{x \mid x \in E \wedge \|x\| \leq 1\}$ and $S^{\infty} := \partial B^{\infty}$. Does there exist a retraction $r: B^{\infty} \to S^{\infty}$ which is also a 1-set-contraction? In this connection it is well-known (and trivial) that a nonexpansive retraction $r: B^{\infty} \to S^{\infty}$ does not exist. Moreover there are some other indications inducing that the answer to this question seems to be "no". The following problem is a classical one:

8. Let $(E, \|\ \|)$ be a strictly convex reflexive Banach space and let $\emptyset \neq X \subset E$ be a weakly compact convex subset. Let $f: X \to X$ be nonexpansive. Does f have a fixed point? For isometries it is true that f has a fixed point, [18]. Under a mild compactness condition it is true in general, see [4].

9. Let $(E, \|\ \|)$ be a uniformly-convex Banach space and let $\emptyset \neq X \subset E$ be a closed bounded starshaped subset. Let $f: X \to E$ be nonexpansive with $f[\partial X] \subset X$. Does f have a fixed point? For Hilbert spaces the answer is yes, see [20]. For the spaces l_p $(1 < p < \infty)$ a somewhat weaker result is valid, see [19].

10. Let $(E, \|\ \|)$ be a uniformly-convex Banach space and let $\emptyset \neq X \subset E$ be an open bounded symmetric neighborhood of the origin. Let $f: \bar{X} \to E$ be nonexpansive such that $f(-x) = -f(x)$ for $x \in \partial X$. Does f have a fixed point? Again, for Hilbert spaces it can be shown that f has a fixed point, see [20], [21] and the same can be said - in a limited sense - for the spaces l_p $(1 < p < \infty)$, see [19].

11. Let $(E, \|\ \|)$ be a reflexive Banach space with normal structure. Let $X \subset E$ be an open bounded neighborhood of the origin. Let $f: \bar{X} \to E$ be nonexpansive such that $\forall_{x \in \partial X} \ \forall_{\lambda \in \mathbb{R}} \ f(x) = \lambda x \Rightarrow \lambda \leq 1$. Does f have a fixed point? For Hilbert spaces the answer is yes, see [21]. If either X is assumed to be convex or $(E, \|\ \|)$ is assumed to be uniformly-convex, there are partial results, see [1], [8], [15].

12. Let $(E, \|\ \|)$ be a Banach space and let $\emptyset \neq X \subset E$ be a weakly compact convex subset. Let further $f, g: X \to E$ be such that

(i) f is a generalized contraction in the sense of W.A. Kirk, see [9], [10],

(ii) g is a compact map,

(iii) $(f+g)[X] \subset X$.

Does $f+g$ have a fixed point? f has a fixed point if $g = 0$ (even in a more general setting, see the next problem) or if (iii) is replaced by the much stronger Krasnoselsky-condition (see [12]):

(iv) $\forall_{x,y \in X} \; f(x)+g(y) \in X$ (take $E := \mathbb{R}$, $X := [0,1]$, $f := \frac{1}{2} I$, $g := 1-f$).

Also, the answer is yes if in addition one of the following conditions is assumed to be fulfilled:

(α) $(E, \|\ \|)$ is uniformly-convex,

(β) $(E, \|\ \|)$ satisfies a weak Opial condition (see [22]),

(γ) f is weakly continuous (see [22]).

Finally:

13. Let $(E, \|\ \|)$ be a Banach space and let $\emptyset \neq X \subset E$ be weakly compact; let $f: X \to E$ be a generalized contraction (in the sense of W.A. Kirk, see above) with $f[\partial X] \subset X$. Does f have a fixed point? It is true that f does have a fixed point if $f[X] \subset X$ or if X is assumed to be convex or if $\bigcap_{n \in \mathbb{N}} \operatorname{dom}(f^n) \neq \emptyset$, see [22].

Note (added June 1976):

(a) G. Müller has solved problem 4 affirmatively and problem 5 negatively.

(b) Ten open problems are discussed in a recent paper of S. Reich [16].

References

[1] Browder, F.E., Semicontractive and semiaccretive nonlinear mappings in Banach spaces, Bull. Amer. Math. Soc. 74 (1968) 660-665.

[2] Browder, F.E., Normal solvability and ϕ-accretive mappings in Banach spaces, Bull. Amer. Math. Soc. 78 (1972) 186-192.

[3] Browder, F.E., Asymptotic fixed point theorems, Math. Ann. 185 (1970) 38-60.

[4] Darbo, G., Punti uniti in transformazioni a condominio non compatto, Rend. Sem. Mat. Univ. Padova 24 (1955) 84-92.

[5] Duffin, R.J., Vibrations of a beaded string analyzed topologically, Arch. Rational Mech. Anal. 56 (1974) 287-293.

[6] Edelstein, M., On nonexpansive mappings in Banach spaces, Proc. Cambridge Phil. Soc., 60 (1964) 439-447.

[7] Furi, M. and Vignoli, A., On α-nonexpansive mappings and fixed points, Accad. Naz. Lincei 48 (1970) 195-198.

[8] Gatica, J. and W.A. Kirk, Fixed point theorems for lipschitzian pseudocontractive mappings, Proc. Amer. Math. Soc. 36 (1972) 111-115.

[9] Kirk, W.A., Mappings of generalized contractive type, J. Math. Anal. Appl. 32 (1970) 567-572.

[10] Kirk, W.A., Fixed point theorems for nonlinear nonexpansive and generalized contraction mappings, Pacific J. Math. 38 (1971) 89-94.

[11] Klee, V., Shrinkable neighborhoods in Hausdorff linear spaces, Math. Ann. 141 (1960) 281-285.

[12] Krasnoselsky, M.A., Two remarks on the method of successive approximations, Uspehi Mat. Nauk 10, No 1 (63) (1955) 123-127.

[13] Nussbaum, R.D., Some asymptotic fixed point theorems, Trans. Amer. Math. Soc. 171 (1972) 349-375.

[14] Petryshyn, W.V., Remarks on condensing and k-set-contractive mappings, J. Math. Anal. Appl. 39 (1972) 717-741.

[15] Reich, S., Remarks on fixed points II, Rend. Accad. Naz. Lincei 53, (1973) 170-174.

[16] Reich, S., Some fixed point problems, Rend. Accad. Nat. Lincei, 57 (1974) 194-198.

[17] Reinermann, J., Fixpunktsätze vom Krasnoselski-Typ, Math. Z., 119 (1971) 339-344.

[18] Reinermann, J., Neue Existenz- und Konvergenzsätze in der Fixpunkttheorie nichtlinearer Operatoren, J. Approximation Theory, 8 (1973) 387-399.

[19] Reinermann, J., Fixed point theorems for nonexpansive mappings on starshaped domains, Proceedings on a Conference on "Problems in Nonlinear Functional Analysis", Bonn (1974).

[20] Reinermann, J. and V. Stallbohm, Fixed point theorems for compact and nonexpansive mappings on starshaped domains, Comment. Math. Univ. Carolinae 4, (1974) 775-779; Math. Balkanica 4, (1974) 511-516.

[21] Reinermann, J. and R. Schöneberg, Some results and problems in the fixed point theory for nonexpansive and pseudocontractive mappings in Hilbert space, this volume, 187-197.

[22] Reinermann, J. and R. Schöneberg, Some new results in the fixed point theory of nonexpansive mappings and generalized contractions, this volume, 175-186.

[23] Sadovsky, B.N., On measure of noncompactness and condensing operators, Probl. Math. Anal. Sloz. Sistem. 2 (1968) 89-119.

[24] Sieklucki, K., On a class of plane acyclic continua with the fixed point property, Fund. Math. 63 (1968) 257-278.

Index

A 6
B 7
C 8
D 9
E 0
F 1
G 2
H 3
I 4
J 5